Beekeeping
Naturally

A Simple Recipe

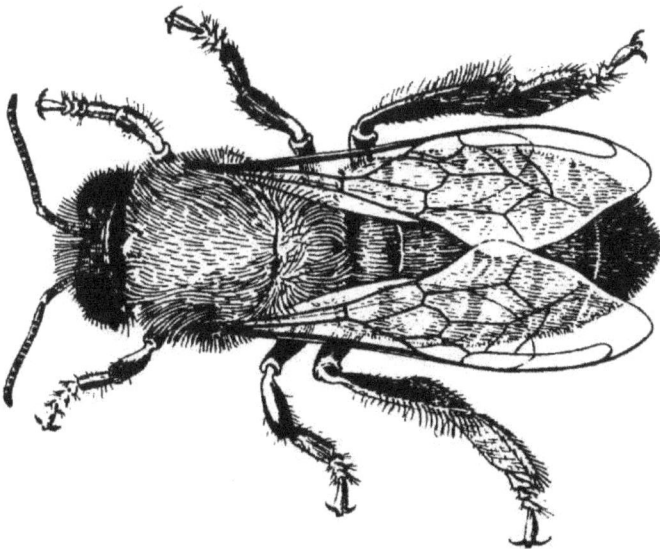

By Michael Bush

Beekeeping Naturally-A Simple Recipe

Cover Photo © 2011 Alex Wild www.alexanderwild.com

ISBN 978-161476-073-3

X-Star Publishing Company
Nehawka, Nebraska, USA
http://xstarpublishing.com

214 pages
75 illustrations

Acknowledgments

Many people helped me along the path of treatment free beekeeping. But none more than the bees. Thank you bees.

Foreword

I hope you will use this book successfully. I tried to make it simple enough but sometimes terminology that seems obvious to me may not be obvious to you. I assume you have some idea of insect biology and some idea of what beekeeping equipment is. If I use any terminology or acronyms that are unfamiliar there is a glossary of terminology and of acronyms in the appendices. The methods I put forth here have worked well for thousands of people who have given me that feedback. Very few people have told me that they have failed at it and often that is only the first year or two and after that things went much better. If you look at the statistics of treating and not treating in the surveys there isn't a lot of difference. I do try to prepare you for the possibility of colonies dying. Statically a lot of bee colonies die every year whether they are treated or not. I hope this is not too discouraging. The first year is always the hardest both because the bees are not established yet and because you have no experience to fall back on. Generally as your skills increase so does your success. Also as you have more resources to work with such as drawn comb, the more success you have. I wish you good success.

Honey bee and Bumble bee

Table of Contents

What this book is not

My previous book, *The Practical Beekeeper*, offers a lot of explanations, options and details. This book instead will focus on what I would choose. It is not a book of a lot of options and explanations. It is also not a complete book on beekeeping; it is just what you need to know to avoid a lot of changes later and to get you started and through the first year. It is also not a book on beekeeping biology. There are many books with that information and I highly recommend any prospective beekeeper study some bee biology eventually. But that is not what this book is about. If you would like more information on any of the choices I am suggesting here you can look on my website for the page on that topic or in the appropriate chapter in my larger book, *The Practical Beekeeper*. You will find beekeepers are often very opinionated on the only right way to do something. I do not wish to come across with that kind of attitude, it's just that a lot of people have asked for a simple system instead of a lot of options and that is what this book is.

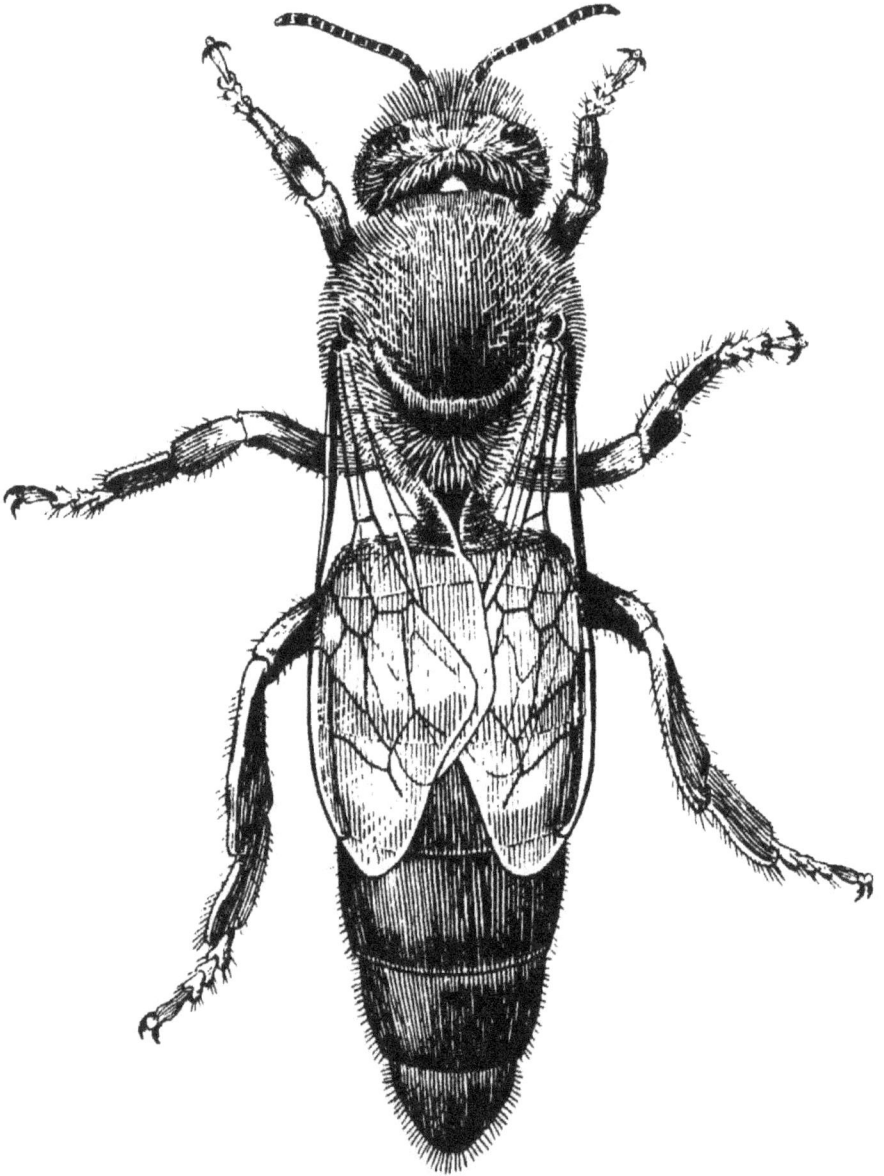

Queen bee

What this book is

This book is a simple recipe for beekeeping naturally. Rather than offer all possible choices I am offering a recipe based on what I would (and have) chosen to use. There are few combinations of equipment I have not used and this is what I arrived at after more than four decades of beekeeping. This is a simple list of things to buy, things to do that worked for me. You will probably encounter people who agree with me and also people who think these choices are wrong. Welcome to beekeeping where for every question you ask of 10 beekeepers you will get 20 answers.

Again, for more detail on many other topics and reasons for my choices, see my web site, www.bushfarms.com or my larger book, *The Practical Beekeeper* available from any online bookseller.

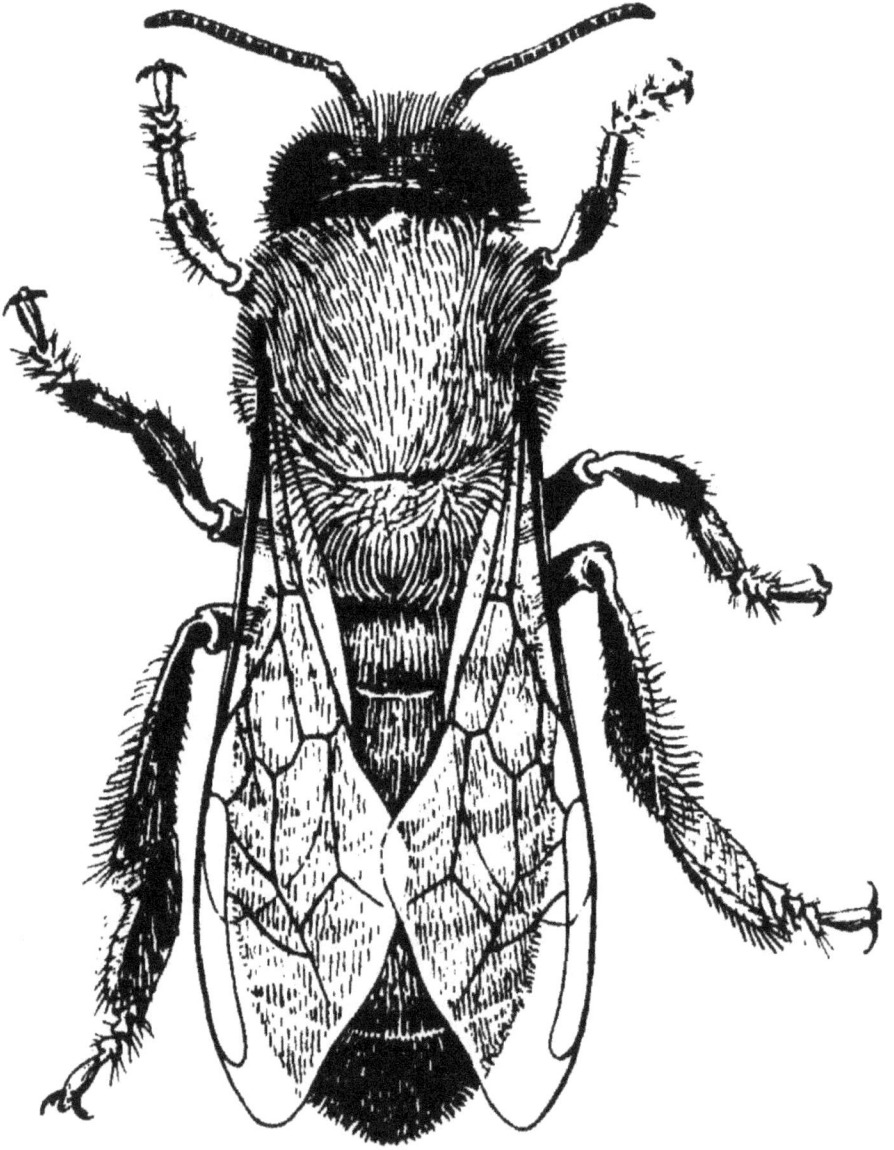

Worker bee

Learning

Newcomers in any field always seem to feel a bit overwhelmed, so before we get too far into this, let's talk about learning.

The most important thing you can learn in life is how to learn. I teach computer classes often and have always been a learner myself. I love to learn. I have discovered, though, that most people don't know how to learn. Here are some rules about learning that I don't think most people know.

"Failure is the key to success; each mistake teaches us something."--Morihei Ueshiba
"A person who never made a mistake never tried anything new." - Albert Einstein

Rule 1: If you're not making mistakes, you're not learning anything. I had a boss in construction who liked to say "If you're not making mistakes you're not doing anything." That may be true, but sometimes you are doing repetitious things and you can get to the point that you are not making mistakes, but if you are learning you will make mistakes! This is a fact. Making mistakes and learning are inseparable. If you're not making mistakes you're not pushing the limits of what you know, and if you're not pushing those limits, you're not learning.

My students in my computer classes often comment on how their children learn computers so quickly and easily and wish it was that easy for them. I tell them why it is easy for children. They are not afraid to make mistakes. Children are used to making mistakes. Adults are not. If you want to learn, get used to making mistakes. Learn from them.

I heard a story about a young man who was taking over as a bank president. The person who held the job before had been there for forty years and had made the company a lot of money. The young man asked him for advice before he left. The old man said that to make the bank money you make good decisions. The young man asked "how do you make good decisions?" The old man said, "You make bad decisions and learn from them." In the end, this is the really the *only* way to learn. Make mistakes and learn from them. I'm not saying you can't learn from other people's mistakes or from books, but in the end you have to make your own mistakes.

"Not understanding is the beginning of Learning. Confusion is the reflection of growth."--Robin Sharma

Rule 2: If you're not confused, you're not learning anything. If you are going to be a learner you will have to get used to being confused. Confusion is the feeling you get when you are trying to figure things out. Adults find this disconcerting, but there is no other way to learn. If you think back to the last card game you learned, you were told the rules, which you couldn't remember, but you started playing anyway. The first few hands were terrible, but then you started to understand the rules. But that was only the beginning. Then you played until you started to understand how to play strategically, but until you got good at it you were still confused. Gradually the whole picture of the rules and the strategies and how they fit together started to congeal in your mind and then it made sense. The only way from here to there, though, is that period of confusion.

The problem with learning and our world view is: we think things can be laid out linearly. You learn this fact; add this one and that one and then finally you know all the facts. But reality is not a set of linear facts; it is a set of relationships. It is those relationships and principles that understanding is made up of. It takes a lot of confusion to finally sort out all the relationships. There is no starting and ending point, because it is not a line, it is circles within circles. So you start somewhere and continue until you have the basic relationships.

"Education is not the learning of facts, but the training of minds to think."--Albert Einstein

Rule 3: Real learning is not facts, it is relationships. It's kind of like a jigsaw puzzle. You start somewhere, even though it doesn't look like anything yet. You sort things out by color and pattern and then you start fitting them together. Everything you learn in any subject is part of the whole puzzle and is related to everything else somehow.

"There are not more than five musical notes, yet the combinations of these five give rise to more melodies than can ever be heard.

There are not more than five primary colors, yet in combination they produce more hues than can ever been seen.

There are not more than five cardinal tastes, yet combinations of them yield more flavors than can ever be tasted."— Sun Tzu, The Art of War

The facts are just the pieces of the puzzle. You need them to figure out the relationships, but the pieces themselves don't make any sense until you have them connected. The connectedness of all things is one of the first things you need to learn in order to be able to learn.

Einstein was once asked how many feet are in a mile. Einstein's reply was "I don't know. Why should I fill my brain with facts I can find in two minutes in any standard reference book?"— Albert Einstein

It is much more important to have a few facts and understand the relationships than lots of facts and no relationships. One little part of the puzzle put together is better than more pieces and none of them put together. Knowledge and understanding are not at all related. Don't go for knowledge; go for understanding, and knowledge takes care of itself.

Rule 4: It's not so important what you know as it is that you know how to find out. I read *Tom Brown's Field Guide to Wilderness Survival*. I read survival guides all the time, but they usually frustrate me because they give recipes. Take this and that and do this with it and you have a shelter. The problem is, in real life you usually don't have one of the ingredients. Tom Brown, though, in his chapter on shelter, showed how he *learned* how to build a shelter. Telling you *how* to build a shelter and telling how to *learn* to build a shelter are as different as night and day. What you want to learn in life is not what the answers are, but how to find the answers. If you know that you can adjust to the materials and situations available.

The usual method is to look around and pay attention. Tom Brown learned to build a shelter by watching the squirrels, but he could have watched any animal that needed shelter and learned from them. Watching how other people and animals solve their problems and adapting those solutions is one way to learn.

Swarm in the air

Bee Basics

In order to do beekeeping, you need a basic understanding of their life cycle and their yearly "colony" cycle. You have two levels of organisms—the individual bee (which can't exist as an organism for very long) and the colony superorganism.

Life cycle of a bee

Bees are one of three main castes: queen, worker or drone. The queen is the one bee that reproduces, but even that she can't do by herself. She is the one bee that goes out and mates, during one period of her life, that lasts a few days, and then she lays eggs for the rest of her life. The workers, depending on their age, feed brood, make comb, store honey, clean house, guard the entrance or gather honey, pollen, water or propolis. The drones spend their days flying out to drone congregation areas (DCAs) in the early afternoon and flying home just before dark. They spend their lives in hopes of finding a queen to mate with. So let's follow each cast from egg to death:

Queen

We will start with the queen since she is the most pivotal of any bee because there is generally only one of her. The reasons the bees raise a queen are: queenlessness (emergency), failing queen (supersedure), and swarming (colony reproduction).

Queenlessness

The cells for each appear slightly different or at least occur under different conditions that can be observed. A queenless hive will have no queen that can be found, little open brood and no unhatched eggs. The queen cells resemble a peanut hanging on the side or bottom of a comb. If the queen died or was killed the bees will take young larvae and feed it extensive amounts of Royal Jelly and build a large hanging cell for the larvae.

Supersedure

In supersedure the bees are trying to replace a queen they perceive as failing. She is probably between 2 and 4 years old and not laying as many fertile eggs and not making as much Queen Mandibular Pheromone (QMP). These cells are usually on the face of the comb about $2/3$ of the way up the comb. There are, of course, exceptions. Jay Smith had a queen that was still laying well at 7 years named Alice, but three years seems to be the norm when the bees replace them.

Swarming

Swarm cells are built to facilitate the reproduction of the superorganism. It's how the colony starts new colonies. The swarm cells are usually on the bottom of the frames making up the brood nest. They are usually easy to find by tipping up the brood chamber and examining the bottom of the frames.

The larvae that make a good queen are worker eggs that just hatched, which happens on day $3^1/_2$ from the day the egg was laid. On day 8 (for large cell) or day 7 (for natural sized cells) the cell will be capped. On day 16 (for large cell) or day 15 (for natural sized cells) the queen will usually emerge. On day 22, weather permitting, she may fly. On day 25, weather permitting, she may mate over the next several days. By day 28 we may see eggs from a new fertile queen. From that time on, she will lay eggs (weather and stores permitting) until she fails or swarms to a new location and starts laying there. The queen will live two or three years in the wild, but almost always fails by the third year and is replaced by the workers. In a swarm the old queen leaves with the first (primary) swarm. Virgin queens leave with the subsequent swarms, which are called afterswarms.

Worker

Worker Bee Gathering Propolis

A worker egg starts out the same as a queen egg. It is a fertilized egg. Both are fed royal jelly at first, but the worker gets less and less as it matures. Both hatch on day $3^1/_2$ but the worker develops more slowly. From day 3 $^{1}/_{2}$ until it is capped it is called "open brood". It is not capped until the 9th day (for large cells) or the 8th day (for natural sized cells). From the day it is capped until it emerges it is called "capped brood". It emerges on the 21st day (for large cells) or the 18th or 19th day (for natural sized cells). From when the bees start chewing through the caps until they emerge they are called "emerging brood". After emergence a worker starts its life as a nurse bee, feeding the young larvae (open brood). For those who say that a worker is an incomplete female while a queen is a fully functioning female, consider that only a worker can produce "milk" for the young. Only a worker can feed and care for the young. The queen does not have the right glands to produce food for young, nor the skills to care for them. Neither the worker nor the queen is a "complete mother"; it takes them both to rear the young. Workers and queens are anatomically different in many ways. Only a worker has the hypopharyngeal gland to feed the young. Only a worker has baskets for carrying pollen and propolis. Only a queen can lay fertile eggs. Only a queen can make sufficient pheromones to keep the hive working correctly.

For the first 2 days the newly emerged worker will clean cells and generate heat for the brood nest. The next 3 to 5 days it will feed older larvae. The next 6 to 10 days it will feed young larvae and queens (if there are any). During this period from 1 to 10 days old it is a Nurse Bee. From day 11 to 18 the worker will make honey, not gather but ripen nectar and take it from field bees bringing it back, and build comb. From days 19 to 21 the workers will be ventilation units and guard bees and janitors cleaning up the hive and taking out the trash. From day 11 to 21 they are House Bees. Day 22 to the end of their life they are foragers. Except during winter, workers usually live about six weeks or less, working themselves to death until their wings are too shredded to fly. If the queen fails a worker may develop ovaries and start to lay. Usually these are drone eggs and usually there are several to a cell and they are in worker cells.

Drone

Drones are from unfertilized eggs. For those of you who studied any genetics, they are haploid, meaning they only have a single set of genes, where a worker and a queen are diploid, meaning they have pairs of genes (twice as many). Drones are larger than workers but proportionately wider, shorter than a queen, have a blunt back end, have huge eyes and no stinger. The egg hatches on day $3^1/_2$. The cell is capped on day 10 (for large cells) or as early as day 9 (for natural sized cells) and emerges on day 24 (for large cells) or between day 21 and 24 (for natural sized cells). The colony will raise drones whenever resources are plentiful so that there will be drones to mate with a queen if they are needed. It is unclear what other purposes they serve, but since a typical hive raises 10,000 or more of them in the course of year and only 1 or 2 ever get to mate, they may serve other purposes. If there is a shortage of resources the drones are driven out of the hive and die from cold or starvation. The first few days of their lives they beg food from the nurse bees. The next few days they eat right from the open cells in the brood nest (which is where they usually hang out). After a week or so they start flying and finding their way around. After about two weeks they are regularly flying to DCAs (Drone Congregation Areas) in the early afternoon and stay until evening. These are areas where drones congregate and where the queens go to mate. If a drone is "fortunate" enough to mate, his reward is to have the queen clamp down on his member and rip it out by the roots. He will die from the damage. The queen stores up the sperm in a special receptacle (spermatheca) and distributes it as she lays the eggs. When the queen runs out of stored sperm, she does not mate again, she fails and is replaced. I think drones have an undeserved reputation for being useless. In fact they are essential. Not only do they have a reputation for being useless but for being lazy. They are not lazy. They fly until they are exhausted every day that the weather permits, trying to ensure the continuation of the species.

Yearly cycle of the colony

By definition this is a cycle so we'll start when the year really begins, in the winter. I can speak to what happens in Nebraska. For your location I would consult local beekeepers.

Winter

The colony tries to go into winter with sufficient stores, not only to survive the winter, but to build up enough by spring for the colony to reproduce. To do this the colony needs a good supply of honey and pollen. The bee colony appears to be dormant all winter. They usually don't fly unless the temperatures get up around 50º F (10º C). But actually the bees maintain heat in the cluster all winter and all winter the colony will rear little batches of brood to replenish the supply of young bees. These batches take a lot of energy and the cluster has to stay much warmer during them. The colony takes breaks between batches. As soon as there is any supply of fresh pollen coming in the colony will begin buildup in earnest. Usually the early pollen is the Maples and the Pussy Willows. In my location this is late February or early March. Of course if the weather isn't warm enough to fly, the bees won't have any way to get it. Beekeepers often put pollen patties on at this time so the weather won't be a deciding factor in the buildup.

Spring

By spring the colony is now building up well. They should have raised at least one turnover of brood by now. They will really take off with the first bloom. This is usually dandelions or the early fruit trees. Here in Nebraska, that's the wild plums and chokecherries which will bloom about mid-April. Between now and mid-May the colony will be intent on swarm preparations. They will try to finish building up and then start back filling the brood nest with nectar so the queen can't lay. This sets off a chain reaction that leads to swarming. The more the queen doesn't lay the more she loses weight so she can fly. The less brood there is to care for, the more unemployed nurse bees there are (the ones who will swarm). Once critical mass of unemployed nurse bees is reached, they will build swarm cells, the queen will lay in them and the colony will swarm just before they are capped. All of this is assuming, of course, that there are abundant resources and that the beekeeper doesn't intervene. If they decide not to swarm then they go full throttle into nectar collection. If they decide *to* swarm then the old queen leaves with a large amount of the young bees and try to start a new home somewhere. Meanwhile the new queen emerges in a couple of weeks and starts laying in another couple of weeks and the remaining field bees haul in the crop to build up for the next winter.

Summer

Our flow, here in Nebraska, is really mostly in the summer. This is usually followed by a summer lull. It seems to be driven, here in my location anyway, by a drop in rainfall. Sometimes if the rain is timed right there isn't really a lull at all, but usually there is. Our flow starts about mid-June and ends when things dry up enough. Sometimes there's an actual dearth where there is no nectar at all and the queens stop laying. I'd say most of my nectar is soybeans, alfalfa, clover, and just plain weeds. This varies greatly by climate.

Fall

We usually get a fall flow in Nebraska. It's mostly smartweed, goldenrod, aster and chicory with some sunflower and partridge pea and other weeds. Some years it's enough to make a crop. Some years it's not enough to get them through the winter and I have to feed them. Around mid-October, usually, the queens stop laying and the bees start settling in for the winter.

Belief as relates to beekeeping

"If you think you can or you think you can't, you are right"--Henry Ford

How does belief influence success?

I want to see you succeed and one of the ingredients to success is belief. As soon as you talk about belief affecting results, there is an assumption by some that you are being unscientific yet all the current research on the success or failure of computer projects or any projects in a business environment has established the fact that success or failure is dependent on "buy in" from the employees and from management. Anyone who has observed scientific research can see this effect as well. Whether you want to believe it or not, "buy in" *is* "belief". Believing it can work, and believing it needs to work. There seems to be a group that thinks that belief has nothing to do with, or should have nothing to do with, success or failure, but only the "facts". But all the way from "the little engine that could" to real life stories of success like Edison testing thousands of filaments to come up with a practical light bulb illustrate that belief is equally important to the success of any undertaking.

Please, do not be confused by what I'm saying. I'm not saying that Edison could just "believe" any one of those filaments into working. But he had to believe that there was a filament that would work and that is what drove him to keep trying until he found one that would work. You can work as hard as you can and still fail. You can believe as much as is possible and still fail.

It requires adjusting the details of your attempt to match what you discover about reality along the path. But without belief that what you are attempting is possible and worth doing, you will try one time and give up. Then you will say "we tried that once and it didn't work."

Cognitive Dissonance

I think part of this is that you can't really operate well with a cognitive dissonance between what you are attempting to do and your view of the world. If you are attempting to do something that, in your view of the world, is impossible, it is very unlikely that you will succeed. If you believe there is a solution and you are focused on finding that solution within the framework of your view of the world, you will likely find something that will work. I think in order to succeed at beekeeping; you need to do your beekeeping within the framework of your beliefs.

It's all in the details.

Success and failure of any venture is all in the details. And belief is what drives us to work out the details. I can prove most any controversial beekeeping question in either direction depending on what you want for an outcome. The reason it is controversial and the reason there are two directly opposing beliefs on so many subjects is exactly that—success or failure is dependent, not on the underlying principle being discussed, but the surrounding circumstances. Someone whose experience was under one set of circumstances comes to one conclusion. Someone whose experience is under a different set of circumstances comes to an entirely different conclusion. Change the circumstances and you change the outcome.

Example of details

Let's take it out of the realm of beekeeping. I'll try this two ways, the first is the way it actually happened. A friend called up to tell me that her pressure tank on her well pump was leaking and wanted to know what she should do. I said it was probably one spot in the tank that lost whatever rust proofing the inside of the tank had and it had rusted through. I said I would:

- Buy a fine threaded self-tapping oil plug.
- Buy a bit exactly the size of the shaft (not the threads) of the plug or slightly smaller.
- Buy some gasket sealer.
- Drill out the rust spot.
- Put gasket sealer on the plug.
- Screw the plug in the hole.

When I told her this, I was informed "we tried that and it didn't work". So I went over to her house and looked at her tank. There was a lag screw in the hole... and of course it was still leaking horribly. I then did exactly as I had instructed and fixed it. It lasted at least five or six years after that. Did it work because of my belief? No, but my belief is what gave me the path to success. I believed it was possible and adjusted the details to the reality of the situation.

Now let's try the other way. I also could have just told her to put a bolt in it. It would have been technically correct, but lacking in the details that would actually make it work. But since she ignored the details anyway, I guess it would have worked exactly as poorly. But now let's look at this another way. Why did my "put a bolt in it" work and hers did not? Because I did everything I could to stack the deck in favor of success. Why? Because I believed it could work and therefore I made the effort to make it work. I did not do a halfhearted "we'll try it and see". I went at it from the start with the belief and expectation of succeeding and then doing whatever I could to make what I believed, happen. Then, even if that had failed, I had several backup plans one of which I had used in the past and years later when the threads finally rusted out on the fix on her pressure tank I used it again with success..

My point is that "buy in" *i.e.* "belief" has everything to do with success. Now I will grant that however much she might have believed that lag bolt would stop the leak (and I don't think she really did), it never would have. But if she had focused on how to make it work, she might have succeeded by tweaking the details of the idea to optimize its chances of success and eventually made it work. That tweaking of course improves with experience and sometimes it takes some experimenting to find the right details (Edison and his thousands of filaments). But you can also improve it a lot by listening to someone who has done it before (I've fixed rusted out holes in tanks for decades now and that's how I knew what would have a good chance of working). The same is true of beekeeping.

"People who say it cannot be done should not interrupt those who are doing it."--George Bernard Shaw

---

Rather than work the details out yourself, learn from someone who already has.

Part of belief leading to success is that it gives you the drive to not just give something a cursory try, but to work out the details. And this brings me to another point of frustration. That almost every time someone does an experiment they don't bother to find someone succeeding and ask them about the details before they set up their experiment to prove it doesn't work. If you don't believe in it, you won't make it work. Why not find someone who is succeeding and study them to figure out if it works and then why it works. At that point you will believe it works (because you have observed it) and have an idea how to make it work, by copying someone who has succeeded. For instance, if you want to know something about natural cell size why not talk to someone with hundreds of hives with natural comb in them rather than blunder out on your own? What size the bees build depends on a lot of different things like the time of year, the intended use for the comb etc. So again, I say I can probably get you whatever results you would like because I know what affects it and I can set the stage to get what you want, in this case, larger or smaller.

You can't get the right answer when you are asking the wrong question.

"...unless a distinction can be made rigorous and precise it isn't really a distinction."--Jacques Derrida (1991) *Afterword: Toward An Ethic of Discussion,* published in the English translation of *Limited Inc., pp.123-4, 126*

One of the things I loved in the movie "I Robot" was how often the hologram that is talking to Will Smith says "I'm sorry. My responses are limited. You must ask the right questions." Anytime your question is vague or your criteria are vague your results will be meaningless. Let's try a simple mistake I made myself. When I started out beekeeping, I was too poor to buy any books and a lot of the ones at the library were old ones like

Doolittle and Miller. One of the concepts in those books was "abandonment" as a means of clearing the supers. I was inexperienced and oblivious to when the flows were and when I tried the method, it turned into an unmitigated disaster. Robbing escalated to scary and out of control in a matter of minutes and I fought robbing for weeks after. I was never going to do that again. Rather than believe there might be some value to this method I gave it a one-time try and gave it up.

Then I ran into someone who used the method all the time and when I shared my experience they told me it needs to be done in a flow. Never in a dearth. Now that gap between my experience and what was in the book suddenly closed. I could see how someone would think it was a good method and yet my experience was exactly the opposite. So if my question is just "does the abandonment method work well for clearing supers", I have not asked the right question. It is too vague and my results will not be useful, as other people's results will vary greatly from mine depending on other factors that are not taken into account in my question. I may come to a very distinct and obvious conclusion that is very incorrect. Faith in C.C. Miller or the method might have driven me to ask the right question rather than give up. There are probably two questions I need to answer on the issue of abandonment, in order for my answer to have any meaning:

"Does the abandonment method work well for clearing supers in a dearth?"

"Does the abandonment method work well for clearing supers in a flow?"

(By the way, most beekeeping questions should be asked either about "in a flow" and in a "dearth" or in the "buildup" or in the "wind down".)

If I didn't know enough to formulate those questions based on the fact that currently there are such opposing views on the subject then the question should have been:

"Under what circumstances does the abandonment method work well for clearing supers and under what circumstances does it fail?"

Then I might end up with a useful answer instead of a meaningless one. In my experience, people often end up with an answer that is not useful because they are asking the wrong question.

"All models are wrong, but some are useful" -- George E.P. Box

Paradigm vs Reality

"We don't see the world as it is, we see it as we are"--Anais Nin

So let's tie this back to our "model of the world". Reality is infinitely complex and none of us can actually grasp it, so to solve problems we distill it down to some simplified "model" that we believe includes all the relevant issues. This "model" is the paradigm by which we solve the problem. Let's try a simple practical model. My dad always told me that what it takes to make an internal combustion engine run is: gas, air and spark. If you have these three it should start. This worked most of my life, most of the time, until the problem was a jumped timing chain. At that point I had to expand my "model", my "paradigm" to include timing. I need gas air spark and correct timing. Then when you have a small single piston engine with a broken ring or a bad valve, you may have to expand it to include "compression". Now there are many other things taking place, but that's not the point. The point is we build a model just complex enough to solve the problem because we can't take everything into account. This particular model is just on how to get a motor to start. After that there are other paradigms on how to make it run well. Sometimes we find our paradigm is inadequate for the job and we need to adjust it. Our "model of the world" is never "right", it's never "true"—it's just useful or not useful for the problem at hand. But conversely if we try to solve a problem in a way that is at odds with our model of the world (our personal belief system) we don't really know how to tweak the details to make it work because we are outside the bounds of our paradigm in the unknown. Unless we adjust our paradigm, we probably can't make a solution work that is at odds with our model. In other words if we don't believe in our model we probably can't work out the details of the solution.

I had a boss once who theorized that everyone thinks their idea is best because they thought of it. He didn't mean it facetiously, the reason they thought of it was because they used their model of the world to come up with it, and the reason they like it best, is because it is in harmony with their model of the world. Their solution made sense to them because they arrived at it within that framework. The reason it had a decent chance of success for that person is also that they know how to work in that framework and they have "buy in" to the idea. They "believe" in it because it fits how they think.

Success at anything is much more likely if you are working within your belief system.

Success at anything is also much more likely if you are determined to figure out how to make it work.

Expectations

"Blessed is the man who expects nothing, for he shall never be disappointed"--Alexander Pope

"Demand not that things happen as you wish, but wish them to happen as they do, and you will go on well."—Epictetus, The Discourses

"I have no hopes and therefore I have no fears" -- Reepicheep, in Voyage of the Dawn Treader by C.S. Lewis

I think it's important in every aspect of beekeeping to have realistic expectations. Not to say that those may not be exceeded at times, but also at times they will not be met as both failure and success are dependent on many related variables.

As examples, let's consider some of the variable outcomes.

Honey Crop

Typically people tell beginner beekeepers not to expect a honey crop the first year. This is an attempt to set realistic expectations. However a good package with a good queen in a good year (appropriate amounts of well-timed rainfall and flying weather) may far exceed expectations or may not even get well established. But generally a realistic expectation for the beekeeper is that they should get established enough to get through the winter and maybe make a little honey.

Plastic Foundation

People buy plastic foundation (and other plastic beekeeping equipment such as Honey Super Cell fully drawn comb) and sometimes are very disappointed. The bees typically will hesitate to draw the plastic (or use the Honey Super Cell) and this sets them back a bit. Sometimes the bees will draw a comb between two plastic foundations in order to avoid using it. Sometimes they will build "fins" out from the face of the foundation. None of these are unusual, but they also often draw it pretty well. How well they do depends on a combination of genetics and nectar flow. Many people seeing the hesitation decide never to use plastic again. But actually once the bees use it, comb on plastic foundation or even fully drawn plastic comb is used just like any other comb. The delay at first seems like a big setback, and for a package, perhaps it is, but once you get past it there is no problem getting it used after that.

Wax Foundation

People use wax foundation and often it gets hot and buckles or the bees chew it all up or the bees don't want to draw it and they draw fins or combs between. They do this less with than with plastic, but still sometimes they do. The buckled foundation often gets comb build on it and the comb is a mess. Many people after an experience like this say they will never use wax foundation again. But really that's just how the circumstances went. If you put it in on a good flow the bees would not have chewed it and it would have been drawn before it buckled. My point is that people often have unrealistic expectations and when those are not realized, they are disappointed in the method when it was other circumstances that led to the problems.

Foundationless

Some people use foundationless frames. Many have perfect luck with it but some will have bees that just don't get the concept and build some crossways comb. Since this happens just as often in plastic foundation, and wax foundation that has collapsed or fallen out etc. it would not seem that significant, but if the only experience you have is with the foundationless, you may assume that other methods don't have these problems. But they do. Again, genetics and timing of the flow have a lot to do with success or failure.

Losses

New beekeepers often assume that every hive should live forever and every hive should make it through the winter. Some winters, they do. But most winters kill off at least a few of the hives. Obviously the more hives you have the more you will see this happen. I went years without losing a hive, but I only had a few and I always combined any that were borderline on strength and those were the days before Tracheal mites, Varroa mites, Nosema cerana, small hive beetles, and a host of viruses we now have. Now I have around a hundred hives and try to overwinter a lot of nucs, of marginal strength and there are those many new diseases and pests to stress them out. No winter losses is an unrealistic expectation. But high winter losses are a sign that you must be doing something wrong or the weather did something quirky.

I always try to figure out the cause of winter losses. Often it is starvation from getting stuck on brood. Sometimes with nucs or small clusters it's a hard cold snap (-10 to -30 F) and the cluster just wasn't big enough to keep warm. I always look for dead Varroa. Finding thousands of dead Varroa in the dead bees is usually a good indication that the Varroa were the primary cause of their death. A lack of such evidence is probably good evidence that it was something else.

Again, the point is that sometimes wintering exceeds or falls below even realistic expectations. But it's helpful to start with realistic expectations and work from there. Realistic expectations from healthy hives as far as losses are probably in the 10% range with some years worse and some years better.

Splits

One of the common questions I hear from new beekeepers is "how many splits can I make?" Of course the answer to this is probably the most variable of any except, perhaps, "how much honey will my hive make?" The difference between a good year and a bad year in beekeeping varies far more than 10 fold. I've had years where I got 200 pounds of honey from every hive and years where I harvested nothing and fed 60 pounds of sugar (between spring and fall) to every hive. Splits are similar. Some hives can't be split at all. Some can be split five times in a year. Most can only take one split and still make a decent crop of honey and be well stocked for winter.

How many hives in one place?

Another common question about beekeeping is "how many hives can I put in one place?" With awesome forage (like in the middle of 8000 acres of sweet

clover), and good weather, it may be close to impossible to put too many in one place. With poor forage and
drought, it may be that only a few hives is too many. A
typical number that is thrown about is 20. This is a nice
round number that is applicable as a generality, but to
be realistic it will depend on many things and many of
those things vary from year to year.

   The point of all of this is that results in beekeeping vary dramatically based on what is happening
around the bees as well as things like the time of year,
the way they are cared for and so on. It's very difficult
to predict what the outcomes will actually be, so there
is no point in having too high or low of expectations.
Take things as they come and adjust. Be prepared for
both exceptional success and failure and adjust as you
go.

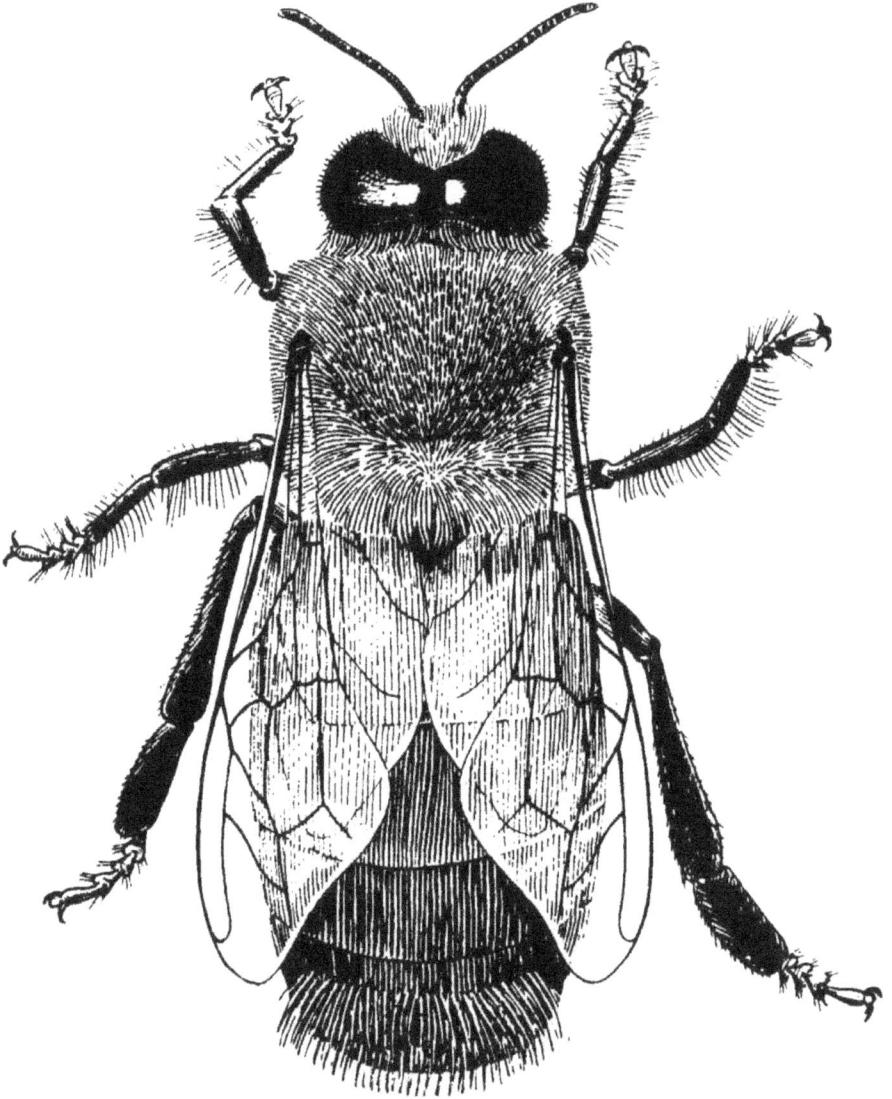

Drone (male) bee

Section 1 Equipment

Basswood or Linden

Boxes: Eight Frame Mediums

Let's establish some terminology. Although bee-keepers often refer to the colony of bees as "the hive", the hive is actually the "house" that the bees live in. The group of bees living in a hive is the colony. There have always been a lot of options as far as equipment. Bee colonies are very adaptable and they can thrive in most any shape hive as long as it's not much too large or much too small. I find it easiest to use standard equipment that is modular in nature. So I find that Langstroth boxes make sense to me. Eight frame mediums make the most sense to me based on things like available frames, foundation, feeders, stability, space management and weight. If you want to look at my reasons I have several web pages on my site or chapters in *The Practical Beekeeper*. There is nothing wrong with other kinds of hives, but there are reasons I have chosen eight frame mediums over those other choices.

Three hives: eight frame, ten frame and eight frame

Frames: Foundationless or PF120s

I will offer two choices in this regard. Foundation is a manmade object intended to guide the bees to build a particular size cell when they build comb. I find that to deal with Varroa I need either natural cell size or small cell size. "Natural" being whatever the bees want to make, which will be smaller than standard foundation, and small cell is 4.9mm cell size. Mann Lake Ltd. has plastic one piece frames that are laid out about 4.9mm and bees given those will build small cells. The other choice would be wooden frames that have no foundation. Also with foundationless, the bees build smaller cells with natural spacing, which is less than standard frames. So if you can get them, a narrow foundationless frame would be the ideal wooden frame. I try to keep track of people who are offering these for sale so check my website for up to date information on availability. Narrow frames would be 1 ¼" in width or 32mm. Standard frames are 1 3/8" in width or 35mm. With narrow frames you can put an extra frame in the box, in other words, you can put 9 frames in an 8 frame box.

Mann Lake PF 120

Narrow (1 ¼″ wide) medium foundationless frame

Bottom Boards: Screened or Solid

Another choice is solid or screened bottom boards. You can buy either, but I would recommend if you get screened bottom boards, leaving a tray under a screened bottom to let the bees control the ventilation. The advantage of the screened bottom is that you can do a certain amount of monitoring of Varroa mites and or other things by pulling out the tray and inspecting it. The solid bottom is simpler and better insulation. Either works fine. I like a top entrance, so I block the bottom entrance with a one by board. A one by two cut the width of the entrance will block the entrance.

Screened Bottom Board

Solid Bottom Board

Covers: Flat or Telescopic

The cover is the roof of the hive. There are many kinds available. My preference is my own homemade covers. Mine are quite simple—just a piece of plywood the size of the box held up enough to make an entrance by shingle shims. If you are somewhat handy they are easy and cheap to make. If you want to buy covers, you can buy either a flat cover or a telescopic cover with an inner cover. To add a top entrance to any cover you can prop it with shingle shims or you can notch it. With a telescopic cover and an inner cover, you can widen the notch or just use the notch if there is one. If I'm using them I like to screen the inner cover hole with #8 hardware cloth and add an empty box on top to get the telescopic cover in the clear for bees to fly directly into the entrance which would be the notch.

Notched inner cover entrance with empty box on top

Top entrance cover with reducer, bottom side

Bottom side of top entrance cover

Top side of top entrance cover

Reduced entrance in use

Feeders

If you are on a budget you can just use an inverted jar with small holes punched in the lid. This can be put over a screened inner cover hole (#8 hardware cloth for screen) or directly on the top bars with something to fill the empty space (rags will work). Never leave empty space or the bees will fill it with comb. Another inexpensive solution for a feeder is a bottom board feeder. Just dam the bottom board with a one by across the entrance, seal up the cracks with wax or chalk, and drill a drain hole for the water to get out when you're not feeding with it.

If you have the money to buy a feeder, I would buy a medium depth frame feeder with cap and ladder for each colony. These are inexpensive and you can leave them in the hive all year around.

This is not to say that there is anything wrong with other kinds of feeders. There are many useful kinds of feeders.

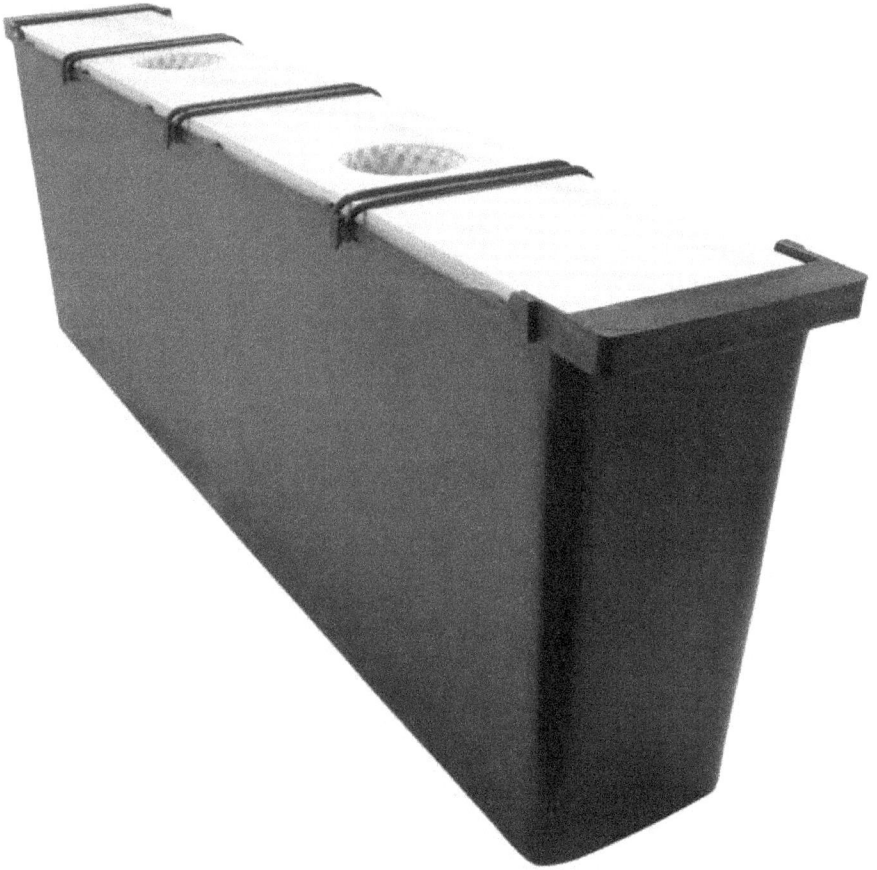

Cap and ladder feeder

How Many Hives and How Much Equipment

It is sound advice to get at least two hives. I think some beginning beekeepers don't get the *purpose* of having two hives, as they often want to experiment with two *different* kinds of hives, like a top bar hive and a Langstroth or a Langstroth eight frame medium and a Langstroth ten frame deep. But this defeats at least one of the purposes of having two hives. A primary reason for having two hives is that the resource that is the hardest to come by and which is often needed to re-solve issues of queenrightness, is frames of brood. But those frames of brood are not of much value if they are not interchangeable. If you really want a top bar hive and a Langstroth hive, then at least make them the same dimensions so the Langstroth frames are inter-changeable with the top bars.

Ideally I would start with two hives but have the equipment for a third hive. The reason is so if you have a swarm or a colony gets too strong and you think you need to split, you have somewhere to put that new colony.

Odds are a new colony won't outgrow five eight frame medium boxes in a year, but it is possible in an awesome year with an awesome queen you might. But again, having enough for three hives improves the odds of not running out. You also need a couple of empty boxes to sort things into when searching for a queen etc. So if we figure five boxes to start, per hive, times three hives is fifteen boxes with frames and two more boxes with no frames. Each hive will require a bottom and a cover and you will need to put that on some kind of stand that won't rot. If we are shaving end bars down to put 9 frames in each box, that means 17 boxes and 135 frames. Having extra frames that aren't assembled yet is fine, but don't assemble them and put them in those empty boxes or you won't have any empty boxes to put frames in while finding a queen or sorting things out.

I would not start with more than four hives, but if you want to start with more than two then do the math and get the extra equipment.

Here are some essentials for the beekeeper:

Large Smoker

I would buy a good smoker. A large one. Large ones are easier to light and keep lit. Smaller ones are harder to light and keep lit. I would light the smoker anytime you are going to do more than just pop the top and I would light it most of the time even then if there is a dearth or any other reason to suspect they might be defensive. Don't over smoke them. Make sure it's lit well and put a puff in the entrance and after you open up a puff across the top bars. Put the smoker down and leave it unless they start to get excited. If you want to see my reasons for smoking and how to smoke, read the chapter by that name.

Some kind of hive tool

Any little flat bar will work. One of my all-time favorites is a very old light cleaver (the blade is about $1^1/_2$" wide and 6" long) that I sharpened on the end. I can pry a box apart or scrape things. It doesn't pull nails well and if the prying is really heavy I do worry about breaking it. If you're going to buy one, I really like the Italian Hive tool I got from Brushy Mountain It's got a lift hook on one end and is light and long has a lot of leverage. But I don't see it in their last catalog. My next favorite is the Thorne hive tool with a frame lifter and next is Maxant's Frame Lifter hive tool. But I do like the Italian one best because the hook fits between the frames more easily.

A bee brush

You can buy one, or if you hunt or have birds you can use a large feather. It has to be a nice stiff quill to do any good. You will need to brush bees off from time to time. In order to harvest, in order to do other manipulations. Shaking can work sometimes, (never when there is a queen cell because you will damage the queen) but sometimes you just need a brush. Like when the bees are all clustering on the edge of the hive you can brush them off before you set the next box on top.

Personal Protective Equipment

Veil, jacket, or suit

Ultrabreeze jacket

I would prefer, if I only have one protective suit, to have a jacket with a zip on veil. It's what I use the most, but it is nice to have a full coverall with a zip on veil. That way I can be pretty fearless of the bees. If you make them mad enough, long enough, they will still get in, but that would require quite a bit of time. If you have the money to spare, I'd buy both. I like the hooded ones, as opposed to the ones with a helmet. I was paranoid at first of the hood being in contact with my head, but I have three nylon outfits (one jacket and two coveralls), and two cotton, all with hoods, and have never been stung on the back of the head like I expected. My favorite jacket is the Ultra Breeze as it is mesh, sting proof and cool on a hot day. It's expensive and worth every penny.

Gloves

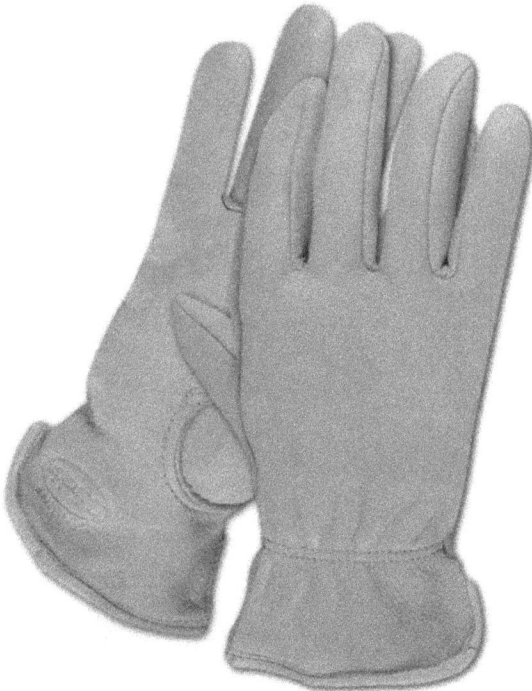

Deerskin gloves

I would wear standard thin leather gloves and tuck them into the sleeves of the jacket. They will be easier to get on and off than the long ones and cheaper to buy.

Parts List

So here is a summary of the minimum to buy to get started with two hives and some extra to handle a swarm or a split:

- 17 eight frame medium boxes
- 135 narrow foundationless frames or 120 Mann Lake plastic PF120 frames or some mixture of the two
- 3 bottom boards
- 3 covers (either migratory or telescopic with inner cover)
- 1 Italian hive tool
- 1 large smoker
- 1 ventilated jacket with zip on veil and elastic or Velcro cuffs
- 1 bee brush (nylon)
- 1 pair of leather gloves (doeskin, deerskin, roping)

Tulip Poplar

Assembling

A lot of equipment is offered assembled these days. That's fine but more expensive and may not be put together as well as you might prefer. You could use screws and glue and they probably won't at the bee supply place. If you choose to glue and screw the boxes, you will need to get deck screws and drill pilot holes so the wood won't split. Frames are much easier to do with a jig. Walter T. Kelley sells a jig. If you buy the PF120s they will not need to be assembled. After you are done the exterior wood (only the outside, not the inside) should be treated in some way to preserve it. Exterior latex paint is popular. If you don't want to cover the pretty grain, you can use tung oil or tung oil and beeswax. If you have some means to heat the boxes without catching them on fire (like a low setting on an oven that never gets over 200 F or so) it will soak in better. If you want to use the mixture of tung oil and beeswax you could heat the tung oil to about 200 F and add the beeswax.

Cleome

Section 2 Bees

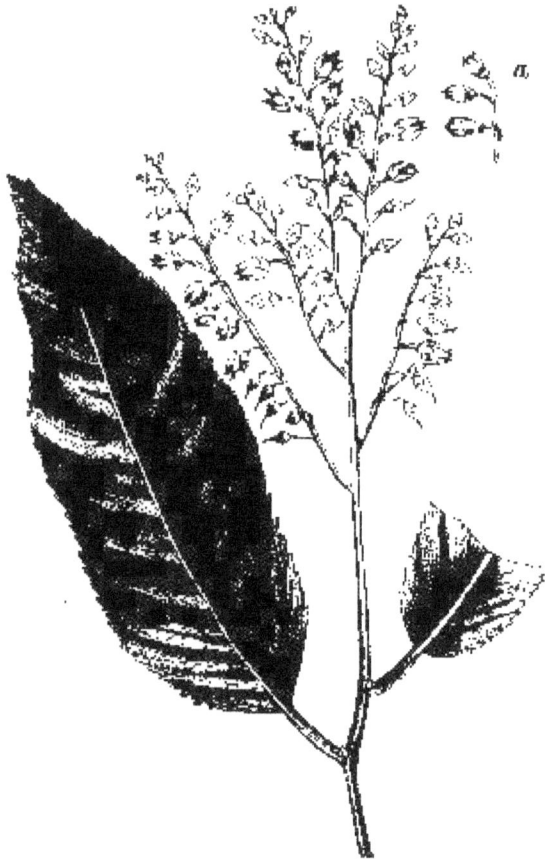

Sourwood

Getting Bees

Most people, treating or not treating, are having very poor luck with packages of large cell treated bees from warmer climates. They tend to do ok up until winter and then they die. I highly recommend trying to find the ideal and then settling for what you can find. Ideally you want small cell, local, treatment free, over-wintered medium depth nucs. Barring that I would try for small cell, local, treatment free packages. Barring that I would settle for as many of those criteria as you can get. If it's not small cell, I'm not so fond of nucs because they will be on comb I don't want and I will be trying to get those combs out, which is another problem. Also if they are on deep frames instead of medium, that introduces another issue.

Swarm

One way to get local untreated bees is to catch swarms of feral bees. If you set out bait hives you might get a swarm to move in, and if they are noticeably small bees, they are probably feral survivor stock. If they are noticeably large bees or they have a marked queen, they are some domestic swarm. Even a local domestic swarm will likely do better than a package from another climate. If it swarmed, it is likely a colony that made it through the winter and so it has decent instincts for your climate. And if it isn't from an over-wintered colony, there is still some chance they will be good enough to survive. If they don't you didn't spend a lot of money for a package that didn't survive.

A bait hive is a topic in itself, but the short version is an empty hive with as much as you can get from the following: A used hive box, lemongrass essential oil (about 4 drops inside and 12 drops on the outside), old brood comb, QMP (queen mandibular pheromone) either as a few drops of the alcohol in which a lot of queens have been left, or the plastic strips sold as pseudoqueen. If you are just starting out you probably won't have the old comb or the used box or the queens in alcohol, but it's worth asking around to see how much of that you can get from some other beekeeper. If not, then use what you can get. Using a bait hive is a good idea. Depending on one is not. If you want fish for supper you might want to go to the grocery store and buy some in the same way if you want bees this year, you may want to buy some. But that doesn't mean you shouldn't "fish" for swarms.

Locating hives

"Where should I put my hive?" The problem is there is neither a simple answer nor a perfect location. But in a list of decreasing importance I would pick these criteria with a willingness to sacrifice the less important ones altogether if they don't work out:

Safety

It's essential to have the hive where they are not a threat to animals who are chained or penned up and can't flee if they are attacked, or where they are likely to be a threat to passersby who don't know there are hives there. If the hive is going to be close to a path that people walk you need to have a fence or something to get the bees up over the people's heads and/or face the hives away from the path. For the safety of the bees they should be where cattle and horses won't rub on them and knock them over, and bears can't get to them.

Convenient access

It's essential to have the hive where the beekeeper can drive right up to it. Carrying full supers that could weigh from 90 pounds (deep) down to 48 pounds (eight frame medium) any distance is too much work. The same reason applies for bringing beekeeping equipment and feed to the hives. You may have to feed as much as 50 pounds or more of syrup to each hive and carrying it any distance is not practical. Also you will learn a lot more about bees with a hive in your backyard than a hive 20 miles away at a friend's house. Also a yard a mile or two from home will get much better care than one 60 miles from home.

Good forage

If you have a lot of options, then go for a place with lots of forage. Sweet clover, alfalfa being grown for seed, tulip poplars etc. can make the difference between bumper crops of 200 pounds or more of honey per hive and barely scraping a living. But keep in mind the bees will not only be foraging the space you own, they will be foraging the 8,000 acres around the hives.

Not in your way

I think it's important the hive does not interfere with anyone's life much. In other words, don't put it right next to a well-used path where, in a dearth and in a bad mood, the bees may harass or sting someone; or anywhere else where you are likely to wish they weren't there.

Full sun

I find hives in full sun have fewer problems with diseases and pests and make more honey. All things being equal, I'd go for full sun. The only advantage to putting them in the shade is that you get to work them in the shade, or it might help meet one of the other more important criteria.

If you live in a very hot climate, mid afternoon shade might be a nice to have, but I wouldn't lose sleep over it unless you have a top bar hive; then I would go for shade to prevent comb collapse.

Not in a low-lying area

I don't care if they are somewhere in the middle between low and high, but I'd rather not have them where the dew and the fog and the cold settle and I really don't want them where I have to move them if there's a threat of a flood.

Out of the wind

It's nice to have them where the cold winter wind doesn't blow on them so hard and the wind is less likely to blow them over or blow off the lids. This isn't my number one requirement, but if a place is available that has a windbreak it's nice. This usually precludes putting them at the very top of a hill.

Water

Bees need water. One of the issues is providing it. Another is to have it more attractive than the neighbor's hot tub. To accomplish this you need to understand that bees are attracted to water because of several things:

- Smell. They can recruit bees to a source that has odor. Chlorine has odor. So does sewage.
- Warmth. Warm water can be taken on even moderately chilly days. Cold water cannot because when the bees get chilled they can't fly home.
- Reliability. Bees prefer a reliable source.
- Accessibility. Bees need to be able to get to the water without falling in. A horse tank or bucket with no floats does not work well. A creek bank provides such access as they can land on the bank and walk up to the water. A barrel or bucket does not unless you provide ladders or floats or both. I use a bucket of water full of old sticks. The bees can land on the stick and climb down to the water.

Conclusion

In the end, bees are very adaptable, so make sure it's convenient for you, and if it's not too hard to provide, try to meet some of the other criteria. It's doubtful you'll have a place that meets all of the criteria listed above.

Installing Package Bees

It occurs to me listening to all of the newbees on the bee forums and watching the YouTube videos of inexperienced people doing their first installs and listening to the experts give advice at new beekeeper classes etc., that there is a lot of very bad advice out there. Sometimes it's just that a beginner doesn't know what a happy medium of something is, but all in all, I think it's just bad advice. So here's my take on a lot of that advice of what to do and not to do:

Not to do:

Don't spray them with syrup

Certainly if you insist on doing this, don't spray them much and don't use thick syrup. 2 parts water to 1 part sugar is plenty. Personally I would not and do not spray them at all. If you have to feed them because you can't get them installed, just spray a little on the screen and wait for them to clean it up. Repeat until they don't take it. But actually I think it's a better plan to refill the can with syrup. Pull it out (of course the bees can now get out so put a board or something over the hole). If you have the kind of can that has a round hole with a rubber grommet holding in a piece of cloth, pop this out and pour in the syrup. Replace the grommet and cloth and then replace the can. If there are just the small holes, then put a hole just big enough for the syrup to run in and fill it full of syrup. Then plug the hole with some softened beeswax. Check for leaks and put the can back.

Why? I've seen many drowned sticky bees from leaky cans or spraying syrup on bees or worse, from overheated bees that regurgitate their honey stomachs as a reflex to cool them off. I don't want to see any more drowned bees. I watched a YouTube video the other day of someone knocking the bees to the bottom (which is fine if you're about to dump them into the hive) soaking them (literally) with syrup, turning the box around and soaking them some more from the other side, then after messing with the hive a bit, soaking them again. I doubt if half of them lived.

I've never seen bees die from *not* spraying them with syrup unless they are starving.

Don't leave them in the shipping box

Don't put them in the hive in the shipping box in order to avoid dumping them out — especially if the box is on top of the top bars with an empty box on top. This is just asking for problems. Assuming you put the queen cage somewhere in the hive, the bees will cluster on the inner cover or cover and then draw combs in the empty box. Bees always prefer their own comb to drawing on foundation and will take every opportunity to do so. Don't give them that opportunity. Bees are not hard to dump out of a box. Yes, this is one of those few things where gentleness and grace are not helpful, but that does not make it hard on the bees or upsetting to the bees. You may as well get used to the idea as someday you'll be shaking a swarm into a box instead of a swarm out of a box. If you really insist on letting them leave the box on their own, then put an empty deep (or medium or whatever) on the *bottom* and put the box in there and then put a box with frames on *top* of that. This takes advantage of the fact that the bees will try to cluster at the top and hang down from there. So hopefully that will be the inner cover and not the bottom bars. Make sure you remove the shipping box and the empty box *the next day*. Not four days later. Not five days later—*the next day*. Otherwise you risk them building comb in the empty space.

Don't hang the queen between the frames

This almost always results in an extra comb between those two frames drawn on the queen cage. Release the queen and you won't have to worry about the messed up combs. This is even more important in a foundationless scenario such as a top bar hive or foundationless frames as one messed up between the frames comb will result in a repeat of the error the rest of the way across. Dump the bees in. Let them settle a bit. To keep the queen from flying, pull the cork from the non-candy end (where she can get out now) and, while holding your thumb over the hole, lay the cage on the bottom and leave it. Put the frames back in and the lid on and walk away. Don't try to release her onto the top bars. Release her down on the bottom board.

One of the issues seems to be that people think that either they will abscond or they will kill the queen. In my experience leaving her caged does not seem to resolve these issues. If they want to leave they usually move to the hive next door anyway and abandon the queen. If you release the queen it also won't stop this from happening, but it also won't cause it. I've not had a problem with a package killing the queen. A bunch of confused bees have been shaken together from many hives and in the confusion they are just happy to find a queen. If they do kill the queen it is almost always because there is already one loose in the package that got shaken in. The bees prefer this queen because they have contact with her.

Don't use an excluder as an includer too long

Don't use an excluder as an includer (to keep the queen *in*) after there is open brood in the hive. I wouldn't use it at all, but there is no point in it after there is open brood and it will keep the drones from being able to fly.

Don't spray the queen with syrup

It will make a mess. Yes, it will probably keep her from flying, but it could also do her harm. I know some think it doesn't but they apparently have not seen a half dead sticky queen before. I've seen plenty. I don't spray her with anything, but if you insist, just use water or at most 2 parts water to 1 part sugar.

Don't install bees without protective equipment

You have enough to worry about without worrying about them stinging you as well.

Don't smoke a package

They are already in a docile mood and they need the pheromones to get organized, find the queen etc. There is no need to interfere with these pheromones as smoking will do little to nothing to calm a swarm or a package anyway.

Don't postpone

Don't postpone installing them because it's a little drizzly or chilly. Unless it's like 10º F or less I would install them and consider it an advantage that they won't want to fly and they will settle in better anyway. Just make sure you have food for them so they don't starve. Capped honey is best. Dry sugar that has been sprayed with enough water to get it damp will do.

Don't feed in a way that makes excessive space

A package is a comb building team. They are looking to build comb everywhere they can. Don't give them space to build it places where they shouldn't. This includes putting empty boxes on top that they have access to, or a spacer for a baggie feeder etc. A frame feeder, a jar over the inner cover with duct tape covering any access or something similar is good. A bottom board feeder is good. Baggie feeders on the bottom board are good *if* you put the bees in first and the baggie feeders on after the bees are off of the bottom.

Don't leave frames out

Ever. Not even for a few minutes. Often you intend to leave them out for a few minutes and forget to come back. When you close a hive up there should always be a full complement of frames in the box, or in the case of a top bar hive, a full complement of bars. Even if you use a follower to temporarily limit the space, fill the empty space with frames or bars. You never know when the bees will find their way over there.

Don't dump bees on top of a baggie feeder

They will get covered in syrup as it all gets squished out by the weight of the bees falling on the baggie.

Don't close up a newly hived package

Let them fly and breathe and get oriented.

Don't leave empty queen cages around

The bees will cluster on them and act like a swarm thinking the cage is a queen because it still smells like one.

Don't let messed up comb lead to more messed up comb

If you have foundationless or a top bar hive this is even more critical. With foundation you get a sort of clean slate every frame as there is another wall of foundation to start from. Still I would try to straighten out any messes quickly. Bees build parallel combs, so with foundationless one bad comb just leads to another. By the same token one good comb leads to another as well. The sooner you make sure the last comb from which the "next" is being built is straight and centered; the better off you will be because the next comb will be parallel to that one. If you have a top bar hive, make sure you have some frames built that you can tie combs into if they get crooked or fall off. That way you can always get at least the last one in the row straight again or, better yet, all of them straight. Especially with foundationless, I would check soon after installation and make sure they are off to the correct start, meaning the combs are in the frames and lined up correctly. The sooner you make sure, the better off you'll be.

If you're using foundation and the bees build fins off of the foundation or parallel combs where there is a gap you can't get to, scrape this off before it has open brood in it. The wax isn't nearly the investment that open brood is. Keep the hive clean of this messed up comb or it will haunt you for a long time to come. With plastic foundation you can just scrape it to the plastic. With wax foundation you'll need more finesse.

Don't destroy supersedure cells

Packages often build supersedure cells and they often tear them back down after a few days, but you tearing them down will risk them ending up queenless. Sometimes there is something wrong with the queen that you don't know. Assuming that the bees are mistaken and you are correct about the quality of the queen is, in my experience, a bad bet.

Don't panic if the queen in the cage is dead

Don't panic and assume they are queenless if the queen in the cage is dead when you get it. Odds are there is a queen loose in the package. Still I would contact the supplier just in case, but meantime install them and come back and check them before you install that new queen. You may just be sentencing her to her doom.

Don't freak out if the queen doesn't lay right away

Some queens will lay as soon as there is comb $1/4''$ deep in the hive. Some take as long as two weeks to start to lay. If they aren't laying in two weeks they probably aren't going to and it's time to freak out.

Don't freak out if one hive is doing better than the other

There are many contributing factors. If they have eggs and brood they are probably doing fine.

Don't get just one hive

Get at least two. You'll have resources then to deal with issues that will come up.

Don't feed constantly

Don't just keep feeding figuring they will stop taking it when they don't need it. I've seen packages that swarmed when they hadn't even finished the first box because they backfilled it all with syrup. Feed until you see some capped stores. This is the sign that the bees have put some of it in "long term storage" meaning they consider it a surplus. If there is a nectar flow at that point, I would stop feeding.

Don't mess with them everyday

They may abscond if you mess with them too often.

Don't leave them on their own for too long

You'll miss the opportunity to learn and you may miss that things are not going correctly. I would check on them within three or four days for the first time and then wait at least that long between visits and try not to go through everything. Just get a general idea how things are going.

Don't smoke them too much

Don't smoke them too much when working them after the install.

Things to do:

Always install them in the minimum amount of space

It takes heat and humidity to raise brood and make wax. Always install them in the minimum amount of space that is large enough and is convenient for you to provide. In other words, if you have a five frame nuc box, that's excellent. If you don't, then use a single box. Yes a single five frame medium box is large enough if you don't have drawn comb in it. An eight frame medium box is large enough if it has drawn comb. While there is nothing wrong, per se, with putting them in more space, in a Northern climate, especially, it is a lot of work for them and they take off much better in a smaller space. While I probably wouldn't *buy* a five frame nuc just for this, I would use it if I had it.

Have your equipment ready

Have your equipment ready before the bees arrive. Have the location picked and the equipment there. Have your protective equipment too.

Wear your protective equipment

You have enough to worry about without thinking about getting stung.

How to install:

When you have everything there, bees, equipment etc., then pull out four or five frames, pull out the can and the queen, slam the box on the ground to knock the bees loose and pour them out like thick oil, or like getting a pick out of a guitar. Tip the box back and forth as needed and when no more will pour slam it again to knock them loose and pour some more. When you are down to ten or twenty bees, set the package down. Pull the cork on a non-candy end (if there is candy) of the queen cage and hold your finger over the hole and set it on the bottom board and let go. Gently set the frames in. Do not push them down on the bees on the bottom. Let the bees move and the frames will settle on their own.

If you release the queen (trickier to make sure she doesn't fly) then do *not* leave the cage. Shake all the bees off of it and put it in a pocket and take it in the house when you are done. Otherwise the bees will cluster on the cage and you'll end up with a queenless swarm on the cage.

Frames tightly together

For some reason this seems to be ignored in the books and causes no end of problems for beekeepers. The frames should be tightly together in the center. A full complement of them (ten for a ten frame box, eight or nine for an eight frame box.). If you leave excess space, the bees are likely to do something funky between, like an extra comb or one out from the face of the comb or fins off of the face of the comb. Your best prevention for this kind of "creative" construction is to push them tightly together. Better yet shave them down to $1^{1}/_{4}$" wide and put an extra frame in tightly together.

Do feed them

A package will go through a lot of feed especially when they have no comb and no stores. Feed them until you start seeing capped honey or they start to backfill the brood nest. Do check on them to make sure things are going correctly. Better to catch things sooner than later, especially things like misdrawn comb.

Section 3 Management

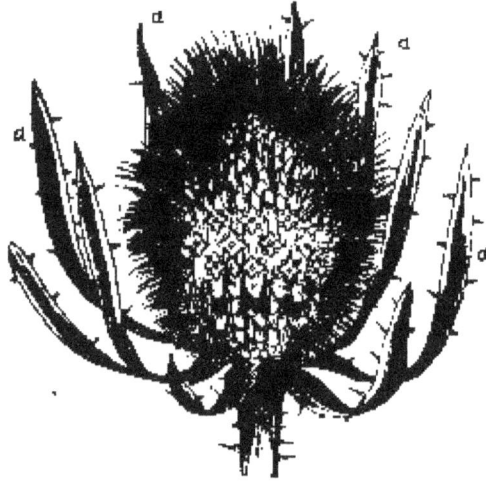

Teasel

Smoke

Don't smoke them too much

Don't smoke them too much when working them after the install. The most common smoking mistakes:

- People have the smoker too hot and burn the bees with the flame thrower they are wielding

- People use far too much smoke causing a general panic instead of simply interfering with the alarm pheromone. One puff in the door is enough. Another on the top if they look excited is ok and after that having it lit and setting nearby is usually sufficient.

- People don't light the smoker because they think smoke upsets the bees, probably because of one of the above reasons.

- People blow the smoke in and immediately open the hive. If you wait a minute the reaction will be completely different. If you're doing something not too time consuming, like filling frame feeders or something, it's a good plan to smoke the next hive before you open this one. That way the minute will be up when you open that one.

- People don't smoke because they have the idea that it is either bad for the bees or somehow unnatural. Their exposure is only a puff or two once every week or two. People have been smoking bees for at least 8,000 years that we have documented for one very good reason. Nothing works better at calming them.

Inspections

First let's talk about how often to inspect.

When you are starting out, you should inspect often because you won't learn much if you don't get into the hive. If you follow my advice and get an observation hive you will not need to open up your hives as often as you can learn from watching them constantly in the observation hive. But assuming you didn't get an observation hive, for learning purposes you probably should get into some hive at least once a week. Probably not all of them. That way you can limit your disruption of the hives to maybe once every two weeks. Probably for your first few years once every two weeks is a good plan. As you get more familiar and better at judging what is happening from the outside of the hive, you can open them less often. So there are two factors here. How to learn about bees without disrupting them too much, and how to manage them (once you have learned a lot) and disrupt them less. Also your personal goals will drive some of how often you need to open the hives. If you are raising queens you will be in the hives more often. If you are just trying to get your garden or your orchard pollinated, you may not need to open them much at all and if you are raising honey, you will need to be somewhere in between those two. Also how often will depend on the time of year. If it's prime swarm season you will need to get into the hives more often. If it's the main flow, you mostly need to keep enough supers on (empty boxes on top for them to put honey in). After the main flow, if you harvest at the normal time, when you don't have a lot of heavy boxes to lift, you can get into them more often with not too much effort and you can keep a better eye on what is happening.   After the flow you'll want to make sure there isn't a dearth.

The next topic is how to do an inspection.

The first step on how is to decide what your goal is.   Possible goals for an inspection: Are they queen-right?   Are things progressing well?   Are the combs getting cross combed or are they straight?   If things don't seem to be strong enough, you might do an extensive inspection to see what is happening.   So once you've established what your goal is, let's do an inspection.

Assuming all you want to do is see if things are ok, you may be able to just pop the top and look in the top box to see if it's being used.   Sometimes, during a flow, all you care is whether or not they need more supers because the existing ones are full.   In this case, a puff of smoke in the door and open the top and take a look.

Assuming you want to be sure they are queen-right then I would do a much more extensive inspection.   Usually you don't need to find a queen, but let's do an extensive inspection where we actually want to find the queen.   Here is the step by step:

First, light the smoker and get it going well.   Everyone has their own methods but here is mine.   First get a large (tall) smoker.   They are much easier to keep lit.   Next pick your fuel.   If you live somewhere that pine straw (pine needles) is abundant and free, it works well enough and smells nice.   But it burns up quickly.   I usually use burlap because it lasts better.   A mixture doesn't hurt.   In other words put some pine straw in the bottom and light that.   Then add a tight roll of burlap on top of that.   It's hard to beat a self-igniting propane torch for lighting it.   I lit my smoker with kitchen matches for decades and it works fine, but it's not as easy as the torch.   Get it going well before doing anything else.   Once you have flames coming out, stop and let it start to smolder.   Then you can put some smoke in the entrance.   It's best to blow some smoke in the entrance and then wait a few minutes.   So if you are doing a row of hives, it's helpful to blow a little into the next hive as well and then as soon as you're closing up the hive, blow a puff into the next hive.

I like to not have to bend over so far, so I usually set an empty box or two on the ground and then a bottom board, or a screened bottom board, or an inner cover or some other board on top of those empty boxes. Then set all of the boxes onto that stack so only a bottom board remains on the stand. Dump off any dead bees, and scrape it down with your hive tool if it needs it. Now set an empty box on the bottom board and start going through the frames in the stack and as you go through them put them in the empty box. The reasons for this are that each of the inspected frames are now in a separate stack. If you were to work your way down the hive without doing this, the queen can be moving around while you are inspecting and you may never see her. Anytime you're going through a hive, keep an eye out for the queen even if finding her is not your goal. It's always good to see her. Also, I like to mark her if I see her. This not only makes it easier to find her next time, but it helps me know if a queen has been replaced. So now you pick up a frame and you search for eggs, larvae and capped brood. Eggs look like a miniature grain of rice standing on end in the bottom of the cell. If there is a dough looking substance in the cell it's probably bee bread which could be anywhere from orange to yellow, to white. Sometimes it's even blue. Nectar looks like water but it's sticky. Capped honey has a waxy cap on it. Sometimes it looks very white because some races of bees leave an air space. These are called "dry cappings". Sometimes it looks more the color of the honey because there is no air space and these are called "wet cappings". Either is fine but it's waxy. Brood, on the other hand, has cappings that look like paper and the color could vary from almost white to dark brown. This is because it has things mixed in so the larvae can breathe. When a package is first installed they mix pollen in. As they raise brood the larvae spins a cocoon and the bees will chew out these cocoons and mix it with the wax to make breathable cappings later.   And so the brood cappings change color over time from light to dark. And comb goes from white, when new, to yellow when finished, to brown when brood has been raised in it, to black eventually.   Worker brood has cappings that are somewhat convex but drone brood is very convex, like Kix cereal.   So you should examine every frame for these things: White wax, bee bread, nectar, honey, eggs, larvae, capped worker brood, capped drone brood, emerging brood (bees chewing their way out of the cells), queen cells (which look like peanuts), workers, festooning workers, drones and the queen.

So what do these things mean?

If you see white wax there is nectar coming in or they would not be building new comb. Also, building new comb is something that happens much more in the spring and is a good indication of a flow. Also festooning bees (bees clinging to each other making curtains of bees in the colony) are indicative of comb building and a healthy density of bees.

If you see a lot of capped worker comb, the population will quickly increase if you see emerging brood the population is quickly increasing. Bee bread, nectar and workers returning with pollen means there are those resources available. Open brood is a good thing, but eggs would be the best evidence that there was a queen at least 3 ½ days ago.

Drone brood and drones means there are plenty of drones to mate with a queen, when there are a lot of them in early spring it's likely that swarm season has started.

Capped honey means they have enough stores to start to put some of it in long term storage.

If you just want to know how they are doing, you may only need to go through the bottom box and then stack them all back up. If you really need to find the queen (which on occasion you do) then keep going a frame at a time until you find her or all of the boxes are back on the original stack. It's handy to put the queen in a hair clip queen catcher so you know where she is. Do *not* put this on top of another hive as those bees may kill her. Keep her on top of one of her boxes.

If it is swarm season, I would usually put some empty frames in the brood nest to keep it open being careful not to spread the bees too thin. They have to keep the brood warm and humid and if you spread them too thin you get stressed bees and possibly chilled brood. My rule of thumb is if you make a space to put your empty frame in and it fills quickly with festooning bees you can add another empty frame.

Space management

Important Concept

Anytime you are inspecting, the space being used and the rate it is being occupied is important. Space management may be the hardest and perhaps one of the most important concepts to grasp in beekeeping. Having the ideal amount of space is a tricky thing to manage but it makes the difference between prosperity and poverty for a honey bee colony. I used to find it very confusing when people would talk about always keeping strong hives. It seemed to me that a nuc or a split was always weak, by definition, but I will offer a new definition. A strong colony is not necessary a colony with a lot of bees, but merely a colony with a good density of bees.

Compression

Once you make this paradigm shift it becomes easier to maintain strong colonies. Any time you see a colony struggling, compress them. Put them in a smaller space. Remove any combs they are not occupying. Remove any combs they have lost control of as evidenced by small hive beetle or wax moth larvae. Freeze those and don't give them back until the colony has grown enough to manage them. An empty foundationless frame is better than an unoccupied drawn comb when you have an issue with hive strength. I call this process "compressing the hive". If you make the hive smaller and increase the density of the bees you will find that a struggling colony is suddenly a booming colony. It's like they were living in a house that had too high of a cost and now they are in one they can afford. Granted they are "affording" it because they have enough bees to do the work, but still they are not over-whelmed by the space that they have to heat, guard and cool. I have seen many a struggling colony turn around quickly when put in the proper space. Slightly crowded is best other than in the main flow when you have to work to keep the space open.

Fear of Swarming

The other issue, of course, is the concern that usually caused the problem of too much space and that is the fear of swarming. Beekeepers often give an over-ly generous space in order to avoid crowding that might cause them to swarm. When you have a colony that is really exploding and the nights are warm and there are plenty of resources coming in, then it's hard to put too many boxes on, but often these are put on too early or left too late. Whenever you see a colony struggling, one of the first things I would do is give them less space.

Here is a study on how much more productive a crowded hive is: Worker-Bee Crowding Affects Brood Production, Honey Production, and Longevity of Honey Bees, John R. Harbo

Nucs

Nucleus hives are handy for this purpose. Eight frame mediums are also handy in that they are half the volume of a ten frame deep, so you have more ability to adjust the space to be "just right". If I have two frames of bees I like to have them in a three frame nuc. If I have six frames of bees I like them in an eight frame box.

Wintering

Winter is another time that just the right space is what you want. I know you'll hear all these people say "the bees don't heat the hive, they just heat the cluster" but I'll guarantee you will be warmer in a small room than a large room when they are cold and both the same temperature. I have spent a lot of my life working outside or semi-outside building houses and little things make a big difference when it comes to cold. I want my hives going into winter with the space they need, not a lot of extra space. Any extra space, if necessary, should be on the bottom. This is part of the concept of overwintering nucs. A small cluster of bees can get through the winter if the density of the bees is high enough.

Aslike Clover

Queen Cells

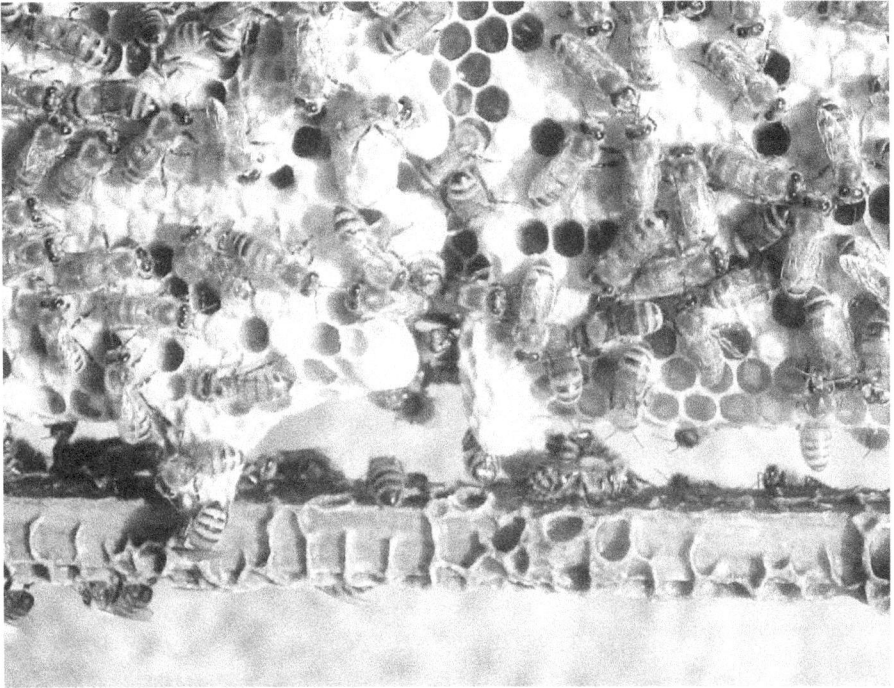

What if you are doing your inspection and you find queen cells? The first thing you need to do is figure out why they are there.

Bees rear queens because of one of four conditions:

Emergency
There is suddenly no queen so a new queen is made from some existing worker larvae.

Supersedure

The bees think the queen is failing so they rear a new one.

Reproductive swarming

The bees decide there are enough bees, and enough stores and enough of the season left to cast a swarm that has a good chance of building up enough to survive the winter without endangering the survival of the colony.

Overcrowding swarm

The bees decide that there are too many bees and not enough room or not enough stores to continue under the current conditions, so they cast an over-crowding swarm as population control. This swarm doesn't have the best chance of survival but the colony believes it improves the colony's chances of survival.

Distinguishing supersedure cells from swarm cells

First, a from swarm cells queen cup with no egg or larva is *not* a queen cell. If there is a larva in it, then it *is* a queen cell.

There is a myth that supersedure cells are always in the middle and swarm cells on the bottom of the frames. This may be a good generality, but you need to look at the entire context of the situation. I would assume that queen cells on the bottom were swarm cells if the hive is building up quickly and is either very strong or very crowded. On the other hand if they are not strong or crowded and building, then I would assume they are not swarm cells. If the cells are more in the middle and conditions otherwise would cause me to expect swarm cells, then I would tend to view these as swarm cells. If the hive were not building and not crowded I would assume they are supersedure cells or emergency cells. Also swarm cells tend to be more numerous.

To put it another way, you should be looking for the underlying causes of swarming. If none of them are present they are probably not swarming, they are probably superseding. If many of those swarm triggers and signs are present, they are probably swarming, not superseding. Those triggers and signs include the time of year, crowding, rapid growth, lots of drone brood and backfilling of the brood nest.

Also swarm cells are not all the same stage of development. Supersedure cells usually are the same age. Emergency cells are usually all the same age. Different ages would be where some have very young larva, some have older larva and some are capped.

What to do

If you think they are swarming and they are already making queen cells, I would split the colony. The bees have decided to start a new colony, so I would do that for them so they don't end up in the trees instead of my hives. If you want a lot of queens you can do a lot of splits from one swarming colony.

If you think it is a supersedure, let them. If you think they are queenless, let them raise their queen.

What not to do

What you *don't* do is tear down the queen cells. It seems like most of the books I've read convince beginning beekeepers that queen cells should always be destroyed. These authors think the bees are either going to swarm, and you want to stop them, or they are trying to replace that precious store-bought queen with a queen of unknown lineage mated with those awful feral drones (in their view). Most of the time when you destroy queen cells the bees swarm anyway, or they already swarmed before you destroyed them, and they not only swarm, but also end up queenless. I see swarm cells as free queens of the highest quality. I put each frame that has queen cells on it, in its own nuc. Usually I try to leave one with the original hive and the old queen in a nuc. That way I've made a bunch of small splits and left the hive thinking it's swarmed already. With supersedure cells, I leave them because the bees apparently have found the queen wanting and I trust the bees. Destroying a supersedure cell is also likely to leave them queenless. The queen is probably about to fail, or she's already failed or died and you just removed their only hope of a queen.

Possible Queenlessness

If you are not sure they are queenless, give them a frame of open brood. There are few solutions as universal in their application and their success than adding a frame of open brood and eggs. If this doesn't resolve things the first week, then do it every week for three weeks. It is a virtual panacea for any queen issues. It gives the bees the pheromones to suppress laying workers. It gives them more workers coming in during a period where there is no laying queen. It does not interfere if there is a virgin queen. It gives them the resources to rear a queen. It is virtually foolproof and does not require finding a queen or seeing eggs. If you have any issue with queenrightness, no brood, worried that there is no queen, this is the simple solution that requires no worrying, no waiting, no hoping. You just give them what they need to resolve the situation. If you have any doubts about the queenrightness of a hive, give them some open brood and eggs and sleep well. Repeat once a week for two more weeks if you still aren't sure. By then things will be fine.

If you are afraid of transferring the queen from the queenright hive, because you are not good at finding queens, then shake or brush all the bees off before you give it to them.

If you are concerned about taking eggs from another new package or small colony, keep in mind that bees have little invested in eggs and the queen can lay far more eggs than a small colony can warm, feed and raise. Taking a frame of eggs from a small struggling new hive and swapping it for an empty comb or any drawn comb will have little impact on the donor colony and may save the recipient if they are indeed queen-less. If the recipient didn't need a queen it will fill in the gap while the new queen gets mated and not interfere with things.

It saves a lot of worry and a lot of judging. Instead you can give them the resources and then observe what they do. If they don't raise a queen, there is probably a virgin loose. If they do raise a queen, they obviously didn't have one or the one they had was not sufficient.

Spring

Tied to climate

Next to wintering this seems to be the next biggest topic of discussion. And, next to wintering, this seems to be the most tied to climate. I can really only share with any confidence what I've actually experienced in my climate. Most places I've had bees are similar (cold winters etc.) but some were a bit colder (Laramie) and some a bit drier (Laramie, Brighton and Mitchell). But all in all most of my experience is in either the Panhandle of Nebraska or Southeast Nebraska. So keep that in mind.

Feeding Bees

Spring is a very volatile and unpredictable time here. We could have warm sunny flying weather and tree pollen as early as late February, but sometimes it stays cold until April. Our first actual nectar availability of any size is the early fruit trees somewhere between early and late April, with mid-April being most likely. The thing that seems to set off spring build up the most is pollen. Feeding syrup is iffy at best. If you feed syrup in February or March (if it ever warms up enough to do so since the bees won't take syrup unless the syrup is 50 F or more) and they decide to brood up a lot and we get a hard freeze (subzero would not be unusual around here) then they could die from trying to keep the brood warm. On the other hand if they don't get going before the first nectar flow in mid-April they won't build up enough to make a good crop. I like to just make sure they have pollen and stores. Dry sugar can stave off starvation. If the weather stays warm enough and they are light enough I might try syrup. I would still stick with 2:1 or 5:3 and not 1:1. 1:1 is just too much moisture in the hive and it doesn't keep well. So my main spring management up until the first blooms is to make sure they have pollen and they don't starve from lack of honey. Once the early flow starts, there is no need to feed really, but if it stays rainy for long periods it might pay. My bottom board feeders are easy enough to feed on the fly like this. Just put in the plugs and fill with syrup even if it's raining. It helps to have a cover to keep the rain out of the syrup if it's really pouring, but if it's just drizzling, the 2:1 will work well and even if it gets watered down the bees still seem pretty interested as it gets diluted, all the way up to 1:2 or more.

Swarm Control

The next issue in spring is heading off swarming. Of course you keep enough supers on that they don't run out of room. But in my experience, this alone will not head off swarming. You need some way to convince them that swarm preparation is not what is happening. I just try to keep the brood nest open. In my location in Eastern Nebraska, in April, they are usually too small to swarm, but if they get going a lot, I'd put more boxes on. They only seem to swarm in April if they get overcrowded. In May is when I have to deal with swarm prevention. The ideal is to keep them from swarming without splitting so you can have a maximum work force to make honey. In order to do this, I recommend keeping the brood nest open. Checkerboarding is fine for this, but as I say I don't seem to have the same conditions that lend well to this. So if a hive is getting really booming and strong from about early May on, I open up the brood nest. I do this with empty frames. No foundation. Just empty frames. Put these in the middle of the brood nest and they are quickly drawn and filled with brood. How many will depend on the strength of the hive. But if the nights aren't that chilly anymore and they can easily fill the gap where I intend to put the empty frame with festooning bees, then I can put another in. The maximum, which should only be done on a really strong hive, is an empty frame every other frame. The minimum, other than none, is one frame.

Splits

If you want more bees and honey isn't your prime consideration then do splits. Sometime on some warm days in April I will try to get all the way to the bottom board and clean it off while looking through the hive for brood, eggs, etc. to make sure things are going well. Other than that I just judge the strength and rate at which the population is increasing. Until you get good at judging this at a glance, look for swarm cells. Usually you can tip a box up and find them hanging down from the bottom of the frames. In the long run, this will give you an idea how much critical mass causes them to swarm and you can judge better how much to inter-vene. If you have swarm cells though, you already missed the opportunity for a large crop and now you need to worry about making splits.

Supering

Of course you need to add supers. You don't want to do this when the hive is still struggling and the weather is cold, but once they are building up you need to add them. Doubling the space of the hive is my goal.

If they are two boxes full, then I add two boxes. If they are four boxes full, then I add four boxes. Of course you eventually may, in a bumper crop year, get so tall you can't do this anymore, but it's a good way to try not to run out of room without giving them more room than they can handle.

Swarm control

Photo by Judy Lillie

Swarming is when the old queen and part of the bees leave to start a new colony. Afterswarms are after the old queen has left and there are still too many bees so some of the swarm queens (which are unmated queens) leave with more swarms. Sometimes a colony has a several afterswarms.

Generally swarming is considered a bad thing because you usually lose those bees. But if you catch them it's a bonus because swarms are notorious for building up quickly. The bees are focused on it already and it's in the natural order of things. Back in the days of skeps and box hives it was always considered a good thing. It was a chance to make increase.

Causes of swarming

It's good to realize that swarming is the normal response of a hive to success. It means they are doing well enough to reproduce the hive. It is the natural order of things. However, it is inconvenient for the beekeeper to have them swarm, so let's think about what causes them to want to swarm.

First there are two main types of swarms. There are reproductive swarms and there are overcrowding swarms. There are a variety of pressures that push them toward swarming.

Overcrowding swarm

Since it's the simplest and can happen anytime, let's briefly look at the overcrowding swarm. The factors that seem to contribute are:

No place to put nectar so it gets stored in the brood nest. Prevention: add supers.

Honey or pollen clogging the brood nest so that the queen has nowhere to lay. Prevention: remove combs of honey and add empty frames so that the bees will be occupied drawing wax and the queen will have somewhere to lay and the bees will have more room to cluster in the brood nest.

Too much traffic congesting the brood nest. Prevention: a top entrance will give foragers a way in without going through the brood nest.

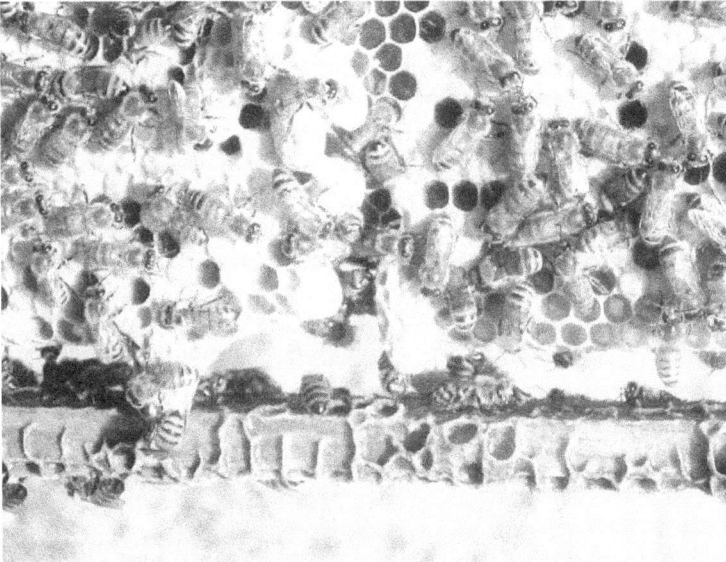

So basically, if you keep supers on and provide ventilation you can prevent an overcrowding swarm.

Reproductive swarm

The bees have been working toward this goal since last winter when they tried to go into winter with excess stores to build up in the spring before the flow enough to afford a swarm that will then have the optimum chance to build up enough to survive the following winter.

The first mistake people make preventing swarms is they think you can just throw on some supers and they won't swarm. But they will. Yes, it's nice to have room for them to store the honey, so the supers are helpful, but the bees intend to swarm and the supers will not deter them from the plan to do a reproductive swarm.

Back to the sequence in the spring, the bees, during winter, rear little spurts of brood. The queen lays a little and they start rearing that batch, but they don't start any new brood until that brood emerges and they take a break. Then they rear another little batch. When pollen starts coming in they start to rear more brood to build up the population of workers. They also start using up the honey they have stored. This is used to feed brood and also it makes room for more brood.

When the bees think they have enough bees they start filling all of that back in with honey, both to stop the queen from laying, and to have adequate stores in case the main flow doesn't pan out. As the brood nest gets backfilled it makes more and more unemployed nurse bees. These nurse bees start doing a keening buzz that is quite different from the typical harmonious buzz you usually hear—more of a warble. Once the brood nest is mostly full of honey they start swarm cells. About the time they get capped the old queen leaves with a large number of bees. Even if you catch the swarm, the hive has still stopped brood production and has lost (to the swarm) a lot of bees. It's doubtful it will make honey. If there are still enough bees, the hive will throw afterswarms with virgin queens heading them.

If I don't catch them in time, once they make up their mind I always make splits because not much will dissuade them. Destroying queen cells only postpones the inevitable, at best and most likely will leave them queenless. My guess is that most people destroy the queen cells *after* the hive has swarmed without realizing it.

I often just put every frame that has some queen cells on it with a frame of honey in a two frame nuc to get good queens.

But, of course, the real object is to avoid the swarm and the split.

Preventing swarming

I do love to catch swarms but who has time to watch the hives all the time to catch them? And if you have that much time, then you have the time to prevent them.

Opening the broodnest

This, of course is what we want to do. What we need to do is interrupt the chain of events. The easiest way is to keep the brood nest open. If you keep the brood nest from backfilling and if you occupy all those unemployed nurse bees then you can change their mind. If you see them starting swarm preparations (backfilling the brood nest) and catch it before they start queen cells, you can put some empty frames in the brood nest. Yes, empty. No foundation. Nothing. Just an empty frame. Just one here and there with two frames of brood between. In other words, you can do something like: BBEBBEBBEB where B is brood comb and E is an empty frame. How many you insert depends on how strong the cluster is. They have to fill all those gaps with bees. The gaps fill with the unemployed nurse bees who begin festooning and building comb. The queen will find the new comb and about the time they get about $^1/_4$" deep, the queen will lay in them. You have now "opened up the brood nest". In one step you have occupied the bees that were preparing to swarm with wax production followed by nursing, you've expanded the brood nest, and you've given the queen a place to lay. If you don't have room to put the empty combs in, then add another brood box and move some brood combs up to that box to make the room to add some to the brood nest. In other words, then the top box would probably be something like EEEBBBEEEE and the bottom one BBEBBEBBEB. It's best to do this as early as they can fill the gap, where you want to put the empty frame, with festooning bees before they start getting "honey bound." The other upside is I get good natural sized brood comb.

A hive that doesn't swarm will produce a *lot* more honey than a hive that swarms.

Splits

What is a split?

A split is sometimes called "making increase". It's when a beekeeper makes one colony into two colonies.

What is the desired outcome?

I would choose my method for doing a split depending on what you want for an outcome.

Reasons for doing a split:

- To get more hives.
- To requeen.
- To prevent swarms.

Timing for doing a split:

As soon as soon as drones are flying and your colony is strong enough, you *can* do a split.

There are an infinite variety of methods for doing a split. Many of these are because of the desired outcome (swarm prevention, maximizing yields, maximizing bees etc.) Some of the variations are also due to buying queens or letting the bees raise queens.

The simple version is to make sure you have some eggs in each of the deeps and put them facing toward the old location. In other words put a bottom board on the left facing the left side of the hive and one on the right facing the right side of the hive and put one deep on each and maybe an empty deep on top of that. Put the tops on and walk away. There are an infinite number of variations of this.

The concepts of splits are:

- You have to make sure that both of the resulting colonies have a queen or the resources to make one (eggs or larvae that just hatched from the egg, drones flying, pollen and honey, plenty of nurse bees).

- You have to make sure that both of the resulting colonies get an adequate supply of honey and pollen to feed the brood and themselves.

- You have to make sure you account for drift back to the original site and insure that both resulting colonies have enough population of bees to care for the brood and the hive they have.

- You need to respect the natural structure of the brood nest. In other words, brood combs belong together. Drone brood goes on the outside edge of the brood and pollen and honey go outside that.

- You need to allow enough time at the end of the season for them to build up for winter in your location.

- The old adage is that you can try to raise more bees or more honey. If you want both, then you can try to maximize honey in the old location and bees in the new split. Otherwise most splits are either a small nuc made up from just enough to get it started, or an even split.

- The size greatly impacts how quickly it builds from there. You can make a split as small as a frame of brood and a frame of honey. But you can't expect that to raise a well-nourished queen. You also can't expect that nuc to build up to a hive by winter. But it makes a good mating nuc or a good place to hold a queen for a while. On the other hand you can make a split that is a minimum of 10 deep frames of bees and brood and honey or 16 medium frames of bees and brood and honey and it will build up rapidly because it has enough "income" and workers to cover its overhead and make a good "profit". They are at "critical mass" and can really grow rapidly rather than struggle to get by. It's more productive and will build up more quickly to do a strong split, let both double in size and do another strong split than to do four weak splits and wait for them to build up.

Split Terminology

These are just individual concepts more so than a particular split.

An even split.

The concept is that the results will be two somewhat equal hives as a result. You take half of everything and divide it up. If you face both of new hives at the sides of the old hive so the returning bees aren't sure which one to come back to. In a week or so, swap places to equalize the drift to the one with the queen.

A walk away split.

The concept here is just that you don't add a queen, you let the queenless colony raise a new one. Make your split but don't add a queen anywhere. Do something to make up for drift. Usually I shake in some extra nurse bees (making sure you don't get the queen), put the lid on and walk away. Come back in four weeks and see if the queen is laying.

Fly back split.

This is a new term I've only seen in since about 2018. The concept is simply that you do a split and the foragers all fly back to the old location.

Swarm control split.

Ideally you want to prevent swarming and not have to split. But if there are queen cells I usually put every frame with any queen cells in its own nuc with a frame of honey and let them rear a queen. This usually relieves the pressure to swarm and gives me very nice queens. But even better, put the old queen in a nuc with a frame of brood and a frame of honey and leave one frame with queen cells at the old hive to simulate a swarm. Many bees are now gone and so is the old queen. Some people do the other kinds of splits (even walk away etc.) in order to prevent swarming. I think it's better to just keep the brood nest open and try to avoid splitting.

Frequently Asked Questions about splits

How early can I do a split?

It's very difficult for a split to build up unless it has an adequate number of bees to keep the brood warm and reach critical mass of workers to handle the overhead of a hive. For mediums this is usually sixteen medium frames of bees with ten of them brood and six of them honey/pollen. I'd say you can split as early as you can put together nucs that are this strong. Half this size can work but a stronger split will take off better. Later in the year when it's not frosting occasionally at night, you could get by with somewhat less, but you'll still do better with this much.

How many times can I split?

Some hives you can't do any splits as they are struggling and never get on their feet. Some hives are such boomers that you can do five splits in a year, although you probably won't get a honey crop.

The object shouldn't be how many can you make, but to keep all the splits you make at critical mass. Critical mass is that point where they are no longer living hand to mouth and they have enough stores, workers, nurse bees and brood to have a surplus. Think of it as economics. If you have barely enough money to pay your bills (or even fall behind on them) you are struggling. When you get to the point where you can pay your bills, you can start to get ahead. When you get to the point where you have some money in the bank and you have a surplus of cash, then life gets pretty easy. Prosperity tends to lead to more prosperity as you can now do things right instead of just getting by. Try it another way. If you run a store you're not really coming out ahead until you cover your overhead.

A hive needs a certain amount of workers to feed the brood (it takes a lot of nurse bees to keep up with a prolific queen), haul the water, pollen, propolis and nectar to feed the brood, build the comb, guard the nest from ants and hive beetles, guard the entrance from skunks and mice and hornets etc.

Once that overhead has been met they can start working on a surplus. If your splits are strong enough to meet their overhead they can take off quickly. If they have barely the resources and workers to survive, they will struggle and take a long time to start really building up.

If you make strong splits and you don't weaken your hives too much you have a shot at getting more splits because they grow faster and more efficiently. Also if you don't weaken your main hives you have more surplus bees to make a surplus crop.

If you take only a frame of brood from each of your strong hives every week they will tend to just make up the difference very quickly with hardly a noticeable lull. One frame of brood and one of honey from each hive put together to fill a ten frame box has a good chance of taking off quickly as opposed to only a few frames of bees.

How late can I do a split?

What you really need to ask yourself is "when is the best time to do a split". By the bee's example that would be sometime before the main flow so they have a flow to get established on. However this tends to cut into your harvest, so you could do them right after the main flow and probably still have time to build up for the fall, if you make them strong enough and give them a mated queen. Of course this depends on the typical flow where you are. If you typically have a dearth after the flow, you may have to feed if you do this

I'm in Nehawka, Nebraska. In a year with a good fall flow, I can do a split on the 1st of August that may build up enough to overwinter in one or two eight frame medium boxes. But if the fall flow fails they may not build up at all.

How far?

The question often seems to come up, how far away to put the split. Mine are usually touching. You need to account for drift if it is less than 2 miles. I've been beekeeping since about 1974 and I've never taken a split 2 miles away unless that's where I wanted to take them anyway. I just do the split and shake in some extra bees or do the split and face both hives to the old location. In other words where the old hive was is where both of the new hives face. Returning bees have to choose. Sometimes I swap those after a few days if one is a lot stronger. Usually the one with the queen is stronger.

I say all of this, mostly because it's "the right thing to do", but really since I went to eight frame mediums and since I expanded to 200 hives, I just split by the box and I do nothing about drift. I put two bottom boards where ever there is room on the stand and "deal" the boxes like cards. "One for you and one for you". I add as much empty room as I have boxes full of bees (in other words I double their actual space). So if there are three boxes full of bees on each stand I add three empty supers with frames. But these are strong splits from booming hives with at least two eight frame medium boxes full of bees in each resulting hive.

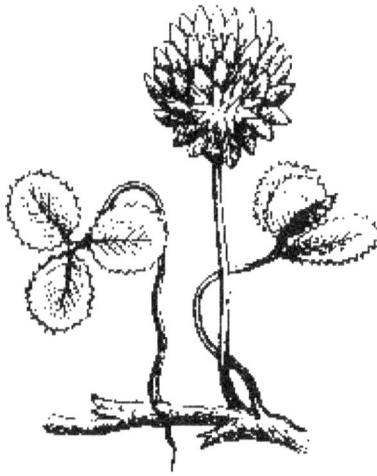

White Clover

Summer

In my climate (and most of the northern U.S.) the main flow is pretty much in the summer. During the main flow the only challenge is to keep them from running out of room. After the main flow the question is whether there will be a dearth and how it will affect things like the buildup or starvation. If you have a dearth it is a bit of a dilemma because feeding often sets off robbing and a dearth makes it much more likely to set off robbing. Yet if you don't feed in a dearth it may set the bees back a lot if they stop rearing brood and the population drops off too much to take advantage of the fall flow. A lot of beekeeping is trying to get a peak population during a flow. In my location that is spring and fall and some years there is a little bit of a flow all summer until the fall flow. Other years there is a dearth in summer. If you feel you need to feed (certainly if the colonies are light on food you may need to), be sure to reduce all the entrances (unless you keep them reduced all the time as I would) and feed all of the colonies, not just the weak ones. If you are going to leave out some colonies, then only feed the strong ones and steal capped honey from them for the weak ones. But it's probably best, if you can do it without setting off robbing, to feed them all so the weak colonies will be more likely to raise more brood.

Borage

Fall

The biggest danger in the fall is that there will be a failed fall flow and that last batch of young "fat" bees won't be raised which are necessary to get through winter. "Fat" bees have more fat bodies which makes them longer lived. They are physiologically different than "skinny" summer bees in this regard. If there is a failed fall flow, I open feed pollen to get that last batch of young bees. To do this I trap pollen during the flow and keep it in the freezer. If I need to feed it (bees gathering sawdust or grain dust is a good indicator of a failed fall flow) then I put down a solid bottom, then a screened bottom then a layer of window screen then an empty box and then put the pollen in the box and top it off with a cover. The window screen keeps the pollen from falling through to the solid bottom. The screened bottom keeps the pollen up where it stays dry so it doesn't mold or turn into small hive beetle food. The bees fly into this otherwise empty hive and gather the pollen just as they do in a flower, by rolling in the pollen pellets and making new pellets.

Ancient Ephesian Honey Bee Coin

Winter

Wintering is so tied to locale that it's hard to know what to write. But I get questions all the time and wish to state what I think on many of the issues. So please read all of this with locale in mind. I will try to cover what I do in my locale (Southeast Nebraska) and why I do it, but that does not mean it is the best for your locale or that some other methods might not work in other or even my location.

I will break this down into topics or manipulations that are commonly discussed whether I do or not.

Another thing that matters is the race or the breed. Mine are all mutts, but they run from brown to black and are Northern bred survivor stock.

I'll break it down by items and actions:

Mouse Guards

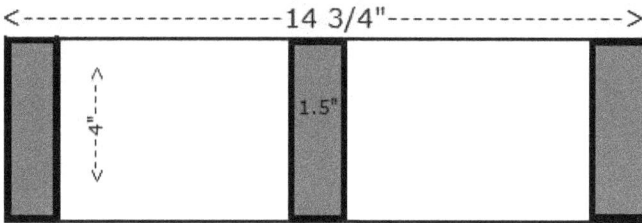

Bottom view of device

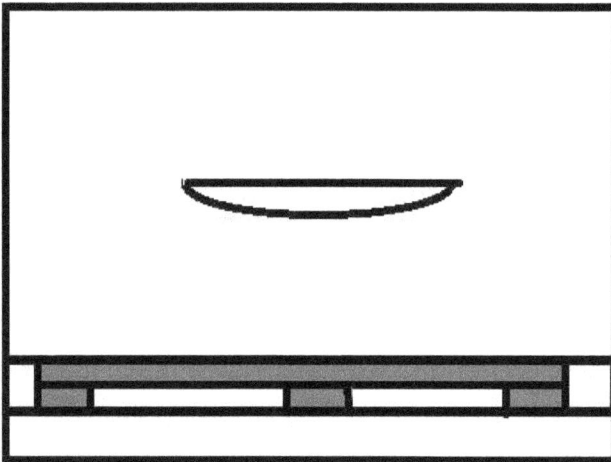

Device in place in the entrance

Typical questions are what to use and when to use them. I have only upper entrances so mouse guards are not an issue. Back when I had lower entrances I used 1/4" hardware cloth for mouse guards, but if I were still using lower entrances I would use a popular device I find here in Southeast Nebraska. The device is a 4" wide piece of 3/8" plywood cut to fit the width of the entrance and three 3/8" laths cut to the 4" width of the plywood. This slides into the entrance reducing it to 3/8" and forming a baffle so that the wind doesn't blow in. People who use it say there is no problem with mice as the 3/8" gap being several inches long seems to deter the mice.

As far as when, with mouse guards I'd try to get them on by or shortly after the first frost. Here we get some warm weather after the first frost, so the mice usually don't move in until it stays cold for several days. You want them on before then or the mice may already be in the hive. The other nice thing about the "baffle" type of entrance reducer/mouse guard is you can leave it in all year around and you don't have to worry about remembering to get the mouse guards on.

Queen Excluders

I don't use excluders, but when I did, I would remove them before winter as they can cause the queen to get stuck below the excluder when the bees move up. The excluder will not stop the bees from moving up, but will keep the queen from joining them. You can store it on top of the inner cover or at the top of the hive I you like, but don't leave it between any boxes.

Screened bottom boards (SBB)

I have these on about half of my hives. If the stand is short enough and enough grass blocks the wind, I sometimes leave out the tray, but usually I put the tray in. Some people in warmer climates seem to think it's good to leave them open year round, but I don't think it works well in a cold windy climate like mine. I also don't think the SBB helps much with Varroa, but it does help with ventilation in the summer and it keeps the bottom board dry in the winter. On the other hand a solid bottom board can double as a feeder and a cover (see "bottom board feeder" in the glossary).

Wrapping

I don't. I tried it once, but it seemed to seal in all the moisture and cause the boxes to remain soaking wet all winter, so I quit doing it.

Clustering hives together

Two stands stacked up

I put my hives on stands that hold two rows of seven (eight frame) hives. Basically they are eight foot long treated two by fours with four foot ends on them. The rails (the eight foot long pieces) are such that the outside ones are 20" from the center and the inside ones are 20" from the outside. This allows the hives (which are 19 7/8") to be all the way forward in the summer to maximize convenience of manipulating them, and all the way back in winter to minimize exposed area. So during the winter 10 of the hives are touching on three sides and the four on the outside ends are touching on two sides. This minimizes exposed walls. Sort of like huddling together for warmth.

Stands with nucs on them.

Feeding Bees

Contrary to popular belief, winter feeding honey or syrup does not work in Northern climates. Once the syrup doesn't make it above 50° F (10° C) during the day (and it takes a while to warm up after a chilly night) the bees won't take it anymore. The time to feed if needed is September and you may be able to continue into October some years. The questions always seem to be what concentration and how much.

When feeding honey, I don't water it down at all. Watered down honey spoils quickly and I can't see wasting it. The best way to feed honey is when it's still in the combs. Feeding it in a feeder often leads to robbing so I usually feed syrup with nothing added that has a smell. When feeding syrup the concentration should not be below 5:3 nor above 2:1. Thicker is better as it will require less evaporation, but I have trouble getting 2:1 to dissolve and if it does it crystallized too fast.

How much is not the right question. The right question is "what is the target weight?" For a large cluster in four medium eight frame boxes (or two ten frame deep boxes) should be between 100 and 150 pounds. In other words if the hive weighs 100 pounds, I might or might not feed, but if it weighs 150 I won't. If it weighs 75 pounds I'll try to feed 75 pounds of honey or syrup. Once the target weight is reached I would stop.

My management plan is to leave them enough honey and steal capped honey from other hives if they are light. But some years when the fall flow fails, I have to feed. I like to wait to harvest until the weather turns cold as it solves several issues. 1) no wax moths to worry about. 2) the bees are clustered below so no bees to remove from supers. 3) I can assess better what to leave and what to take as the fall flow did or did not occur. Another option for a light hive is to feed dry sugar. The down side is that sugar is not stored like syrup, so it's more of an emergency ration, but the up side is you don't have to make syrup, buy feeders, etc. But the sugar not being stored is also the up side. If they don't need it, you don't have syrup stored in your combs. You just put an empty box on the hive with some newspaper on the top bars and pour the sugar on top of the newspaper. I wet it a bit to clump it and wet the edge to get them to see it is food. If the hive is only a little light this is nice insurance. But if it's very light, I think they need to have some capped stores so I'd feed them honey or syrup.

It seems to be a common misconception that feeding syrup can't hurt and people often feed just to be feeding or because it's fall and it's "what you do in the fall". But syrup *does* hurt in several ways. It may be better than starvation, but if they don't *need* it, then it is far better to not feed sugar syrup.

Another issue, while on the subject of feeding with no real purpose in mind, is that people will feed incessantly in the fall until there is nowhere left for the bees to cluster and the humidity in the hive is high from all that syrup that needs to be dried. Then they don't understand why they lose hives in winter.

Bees need to cluster in empty comb where the bees can climb in the cells to compact the cluster. The cluster is often compared to a "ball of bees" but people forget that there is comb between those bees and to be a dense cluster they climb into the cells, which they cannot do if they are filled with syrup.

A solid bottom board can be converted to a feeder (see "bottom board feeder" in the glossary). This makes sense to me because feeding isn't my normal management plan, leaving honey is. Why buy feeders for all your hives if feeding isn't a normal situation? But if I need to feed, I don't have to buy a feeder for each hive. They hold about as much as a frame feeder. Around here candy boards are popular, but the dry sugar on top is easier as you don't have to make the boards, and make the candy. You just use your standard boxes and sugar. I've also sprayed syrup into drawn comb to give a light hive to get them through. It takes the force of the sprayer to get it into the cells. Dipping a comb in syrup does not work.

Insulation

Sometimes I insulate the tops. I gave up insulating anything else. I think it's a good idea to insulate the top, but I just don't always get it done. Since I run a simple top with a top entrance, when I do insulation it's just a piece of Styrofoam on top of the cover with the brick on top of that. This will reduce condensation on the top, as does the top entrance. Any thickness of Styrofoam will do. The main issue is condensation on the lid. When I tried insulating the entire hive the moisture between the insulation and the hive became a problem.

Top Entrances

I think this is essential to reduce condensation in my climate. It was not necessary when I was in Western Nebraska which has much drier climate. It doesn't have to be a large top entrance, just a small one will do. The notch that comes on the notched inner covers is fine. This also provides a way for the bees to exit for cleansing flights on warm snowy days when the bottom entrance (which I don't have) would be blocked with snow. I have only top entrances and no bottom entrances.

Where the cluster is

Usually around here the cluster in the top box going into and coming out of winter, with or without a top entrance. Sometimes it's not, but that seems to be the norm, despite what all the books seem to say. I leave them where they are and I don't try to make them be where I think they should be. Usually they spend the entire winter there.

How strong?

This question comes up a lot. I used to combine weak hives and I seldom lost a hive over winter. However, since I started trying to overwinter nucs I've realized how well a small hive takes off if it does make it through the winter. So I've overwintered much smaller clusters. Also if you have local queens, instead of southern queens, the local queens do better as well as the darker bees overwintering on smaller clusters than the lighter colored bees. So, while I've never seen a softball sized cluster of southern package Italians get through the winter, I've seen that size of feral survivor stock, Carniolans and even Northern raised Italians make it. This is actually going into winter on a cold day (tight cluster). There is some attrition in the fall, and if they are this size in September and there is no flow and they are rearing no brood, they probably are not going to make it. A strong Italian hive going into winter would be a basketball sized cluster or more, while Carniolans or Buckfasts are usually more like soccer ball sized or smaller, and feral survivors tend to be even smaller.

Entrance reducers

I do like them on *all* the hives. On the strong hives they create a traffic jam in the case of a robbing frenzy which will slow things down, and on a weak hive they create a smaller space to guard. On all the hives they create less of a draft than a wide open entrance. In fact when I have forgotten to open up the reducers in the spring, even the strong hives, with the traffic jams because of the small entrance, seem to do better than the ones that are wide open.

Pollen

I have, in recent years, started feeding pollen in the fall during a dearth so they are well stocked with pollen going into winter and so they have one more turnover of brood before winter sets in. There is no point in doing this while real pollen is coming in. I feed real pollen if I have enough. I have sometimes mixed it 50/50 with substitute or soybean flour when I'm desperate and don't have enough. I never mix it at less than 50% real pollen. You can trap this yourself or buy it from one of the suppliers like Brushy Mt. I feed it in the open. I put it on window screen on a SBB on top of a solid bottom board in an empty hive. This would be in September usually.

Windbreak

Some people use straw bales to get a windbreak. I hate mice and they seem to me to be mouse nests waiting to happen, so I don't. But if you kept them back a ways maybe they would work. I suppose one could use corn cribbing or snow fence for a wind break as well as any kind of privacy fence. Mel Disselkoen uses a ring of sheet metal around four hives to make a windbreak for them. This looks like a good setup to me but requires buying the metal and storing it during the rest of the year and then setting it up again in the fall.

Eight frame boxes

I find that eight frame boxes overwinter better than ten frame boxes. The width is more the size of a tree and the size of a cluster, so there is less food left behind. This is not to say that you can't winter bees in ten frame boxes, just that they seem to do slightly better in eight frame boxes.

Medium boxes

I find that medium boxes overwinter better than deeps as there is better communication between frames because of the gap between the boxes. If you picture what is in the hive when the bees cluster in the winter there are combs making walls between parts of the cluster. With a sudden cold snap a group of bees often get trapped on the other side of a deep frame when the cluster contracts as they can't get to the top or bottom and over, where with the medium depth box the cluster usually spans the gap between the boxes providing communication between frames throughout the hive. This is not to say you can't overwinter them in deeps, but only that they seem to do slightly better in mediums.

Narrow frames

I find they winter better on narrow frames (1 1/4" on center instead of the standard 1 3/8" on center or the 9 frame arrangement in a ten frame box which is about 1 1/2" on center) because it takes less bees in the late winter to cover and keep the brood warm than it does with larger gaps. This is not to say you can't overwinter them on 1 3/8" frames, only that they seem to do slightly better, build up earlier, get less chilled brood and less chalkbrood on narrow frames.

Wintering Nucs

I have tried overwintering nucs every winter since 2004. I can't claim to be good at it, but when I get nucs through they are my best hives the next year. I've tried many things from wrapping, huddling, heating, feeding syrup all winter etc. I've come to these conclusions. First, wrapping just made them too wet. Feeding syrup all winter did also. Insulating top and bottom and huddling were helpful. A heater if not too hot, down the middle of this arrangement was helpful, except every year someone unplugs it during the coldest spell, so it really hasn't helped. My nucs are a bit backwards of most in how they come to be, as mine are combines of mating nucs rather than splits from my strong hives or requeening and splits from my weak hives. I've concluded that one mistake I've made is I need to combine them soon enough for them to get reorganized as their own colony before the cold weather sets in. Which means about the end of July or the first of August. This also lets them get some stores put away and arranged the way they want. But assuming you're making splits of your weak hives and requeening them, the same rule holds true. You want them to have time to get organized as a colony. I'm liking the sugar on top more and more for these as feeding syrup has the problem of too much moisture. But if you feed early this isn't so much of a problem. Rather than spend a lot of time making special equipment for overwintering nucs, I think it's more practical to figure out how to overwinter them in your standard equipment. Granted, this makes more sense when your typical box is the size of a five frame deep nuc (my eight frame mediums are exactly that size), but I hate having a lot of specialized equipment around when I can have equipment that is more multi-purpose. My bottom board feeders work well for wintering nucs as you can stack up the nucs and see if they need to be fed and feed any of them without unstacking them.

Indoor wintering

I have not tried it other than the observation hive I typically winter. I have corresponded with many people who have tried it and it is far trickier than one would think. Bees need a cleansing flight now and then so they need to be free flying. They need temps down around 30° to 40° F (-1° to 4° C) to keep them inactive so they don't burn up all their stores and burn out from activity (inactive bees live longer than active bees). Ventilation and keeping bees cool enough seem to be the bigger issues with this than keeping them warm.

Wintering observation hives

I have wintered an observation hive many times. The issues are to make sure they are strong enough going into winter. Have some way to feed them syrup. Have some way to feed them pollen. Don't' over feed the pollen. Make sure they are free flying (check the tube to make sure they haven't clogged it with dead bees and pollen). No, they won't all fly out and die because they are warm and confused about the weather outside. They are quite aware of the weather outside. If they get too weak in the spring you may have to boost them with some bees. A handful or two of bees in an empty box that is connected to the tube will usually result in those bees moving into the hive without you having to take it outside and open it.

Hives clustered together

Conclusion

There are no guarantees of success in beekeeping no matter what method you use. Pick a method that matches your philosophy in life and you will be more successful and happier in your pursuit of beekeeping. Keep in mind that you will get better as a beekeeper the longer you keep bees.

Learn from the bees

"Let the bees tell you"-Brother Adam

I am going to give you the shortcut to success in beekeeping right here. With apologies to C.S. Lewis (who said in *A Horse and His Boy*, *"no one teaches riding quite as well as a horse"*) I think you need to realize that *"no one teaches beekeeping quite as well as bees."* Listen to them and they will teach you.

Trust the Bees

"There are a few rules of thumb that are useful guides. One is that when you are confronted with some problem in the apiary and you do not know what to do, then do nothing. Matters are seldom made worse by doing nothing and are often made much worse by inept intervention." —The How-To-Do-It book of Beekeeping, Richard Taylor

If the question in your mind starts "how do I make the bees …" then you are already thinking wrongly. If your question is "how can I help them with what they are trying to do…" you are on your way to becoming a beekeeper.

Resources

Here, then, is the short answer to every beekeeping issue. *Give them the resources to resolve the problem and let them. If you can't give them the resources, then limit the need for the resources.*

For instance if they are being robbed, what they need is more bees to defend the hive, but if you can't give them that, then reduce the entrance to one bee wide and you will create the "pass at Thermopylae where numbers count for nothing". If they are having wax moth issues in the hive, what they need are more bees to guard the comb. If you can't give them that then reduce the area they need to guard by removing empty combs and empty space.

In other words, give them resources or reduce the need for the resources they don't have.

Panacea

This is so important that I am going to repeat myself so you hopefully get this:

Most bee problems come back to queen issues.

There are few solutions as universal in their application and their success, than adding a frame of open brood from another hive every week for three weeks. It is a virtual panacea for any queen issues. It gives the bees the pheromones to suppress laying workers. It gives them more workers coming in during a period where there is no laying queen. It does not interfere if there is a virgin queen. It gives them the resources to rear a queen. It is virtually foolproof and does not require finding a queen or seeing eggs or accurately diagnosing the problem. If you have any issue with queenrightness, no brood, worried that there is no queen, this is the simple solution that requires no worrying, no waiting, no hoping and no guessing. You just give them what they need to resolve the situation. If you have any doubts about the queenrightness of a hive, give them some open brood and sleep well. Repeat once a week for two more weeks if you still aren't sure. By then things will be well on their way to being fine.

Hive visitor

Appendices

Solitary leaf cutter bee

Glossary

Note: many of these terms are Latin and the plural of the ones with an "a" ending will be "ae". The plural of the "us" endings will be "i". Also meanings are given in the context of beekeeping.

7/11 or Seven/Eleven = Foundation with a cell size that is 700 cells per square decimeter with 11 cells left over. Hence 7/11. Actually 5.6mm cell size. Used because it is a size the queen dislikes laying in because it's too big for worker brood and too small for drone brood. It's only currently available from Walter T. Kelley.

A

Acute Paralysis Virus aka APV = A viral disease of adult bees which affects their ability to use legs or wings normally. It can kill adults and brood.

Abdomen = The posterior or third region of the body of the bee that encloses the honey stomach, stomach, intestines, sting and the reproductive organs.

Abscond = When the entire colony of bees abandons the hive because of pests, disease or other adverse conditions.

Acarapis dorsalis = Mite that lives on honey bees that is indistinguishable from Tracheal mites (Acarapis woodi). It is classified differently simply based on the location where it is found, on the back.

Acarapis externus = Mite that lives on honey bees that is indistinguishable from Tracheal mites. It is classified differently simply based on the location where it is found, on the neck.

Acarapis vagans = Mite that lives on honey bees that is indistinguishable from Tracheal mites. It is classified differently simply based on the location where it is found, anywhere external.

Acarapis woodi = Tracheal Mite, which infests the bees' trachea; sometimes called Acarine Disease or Isle of Wight disease.

Africanized Honey Bees = I have heard these called Apis mellifera scutelata, but Scutelata are actually African bees from the Cape. They used to be called Adansonii, at least that's what Dr. Kerr, who bred them, thought they were. AHB are a mixture of African (Scutelata) and Italian bees. They were created in an attempt to increase production of bees. The USDA bred these at Baton Rouge from stock obtained from Dr. Kerr in Brazil. The USDA shipped these queens to the continental US over the course of many years. The Brazilians also were experimenting with them and the migration of those bees has been followed in the news for some time. They are extremely productive and extremely vigorous bees that are extremely defensive.

Afterswarm = A swarm after the primary swarm. These are headed by a virgin queen.

Alarm pheromone = A chemical (iso-pentyl acetate) substance which smells similar to artificial banana flavoring, released near the worker bee's sting, which alerts the colony to an attack.

Alcohol wash = Putting a cupful of bees in a jar with alcohol to kill the bees and mites so you can count the Varroa mites. A sugar roll is a non-lethal method of doing the same.

Allergic reaction = A systemic reaction to something, such as bee venom, characterized by hives, breathing difficulty, or loss of consciousness. This should be distinguished from a normal reaction to bee venom, which is itching and burning in the general vicinity of the sting.

Alley Method = A graftless method of queen rearing system where bees are put in a "swarm box" to convince them of their queenlessness and a strip of old brood comb is cut and put on a bar for the bees to build into queen cells.

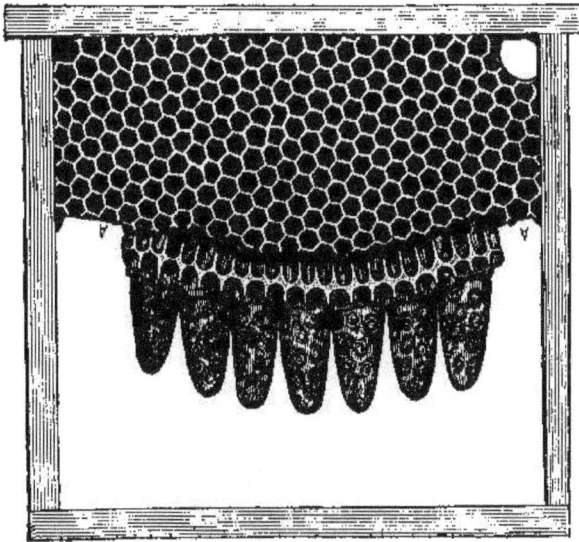

Alley Method

American Foulbrood = For more detail see the chapter on *Enemies of the Bees.* Caused by a spore forming bacteria. It used to be called Bacillus larvae but has recently been renamed Paenibacillus larvae. With American foulbrood the larvae usually dies after it is capped The brood pattern will be spotty. Cappings will be sunken and sometimes pierced. Recently dead larvae will string when poked with a matchstick. The smell is rotten and distinctive. Older dead larvae turn to a scale that the bees cannot remove.

Anaphylactic shock = Constriction of the smooth muscle including the bronchial tubes and blood vessels of a human, caused, in the context of beekeeping, by hypersensitivity to venom possibly resulting in sudden death unless immediate medical attention is received.

Antenna = One of two sensory organs located on the head of the bee, which enable bees to smell and taste.

Attendants = Worker bees that are attending the queen. When used in the context of queens in cages, the workers that are added to the cage to care for the queen.

Apiary = A bee yard.

Apiarist = A beekeeper.

Apiculture = The science and art of raising honey bees.

Apis mellifera mellifera = These are the bees native to England or Germany. They have some of the characteristics of the other dark bees. They tend toward being runny (excitable on the combs) and a bit swarmy, but also seem to be well adapted to damp Northern climates.

Apis mellifera = Includes the honey bees originating in Africa and Europe.

B

Bacillus larvae = The outdated name for Paenibacillus Larvae, the bacteria that causes American foulbrood.

Bacillus thuringiensis = A naturally occurring bacteria that is sprayed on empty comb to kill wax moths. Also sold to control larvae of other specific insects.

Backfilling = A term coined by the late Walt Wright to describe the process of the bees creating a honey bound brood nest. The process where the bees put honey in the brood nest to prevent the queen from laying to prepare for swarming.

Baggie feeder = These are just gallon Ziploc baggies that are filled with three quarts of syrup, laid on the top bars and slit on top with a razor blade with two or three small slits. The bees suck down the syrup until the bag is empty. A box of some kind is required to make room. An upside down Miller feeder or a one by three shim or just any empty super will work. Advantages are the cost (just the cost of the bags) and the bees will work it in cooler weather as the cluster keeps it warm. Disadvantages are you have to disrupt the bees to put new bags on and the old bags are ruined.

Bait hive aka Decoy hive aka Swarm trap = A hive placed to attract stray swarms. Optimum bait hive: At least 20 liters of volume. 9 feet off the ground. Small entrance. Old comb. Lemongrass oil. Queen substance.

Balling = Worker bees surrounding a queen either to confine her because they reject her or to confine her to protect her.

Banking queens = Putting multiple caged queens in one nuc or hive.

Bearding = When bees congregate on the front of the hive.

Bee blower = A gas or electrically driven blower used to blow bees from supers when harvesting.

Bee bread = Fermented pollen stored in the hive to use to feed brood.

Bee brush = Soft brush or whisk or large feather or handful of grass used to remove bees from combs.

Bee escape = A device constructed to permit bees to pass one way, but prevent their return; used to clear bees from supers or other uses. The most common one seems to be the Porter escape which is made to go in the hole in the inner cover. The most effective one seems to be the triangular one which is its own board.

Bee Go = Butyric which is used to drive bees from supers. This smells a lot like vomit.

Bee gum = A piece of a hollow tree used for a hive.

Beehaver = A term coined by George Imirie. One who has bees but has not learned enough technique to be called a beekeeper.

Bee jacket = A white jacket, usually with a zip on veil and elastic at the sleeves and waist, worn as protection when working bees.

Bee Parasitic Mite Syndrome aka Parasitic Mite Syndrome = A set of symptoms that are caused by a major infestation of Varroa mites. Symptoms include the presence of Varroa mites, the presence of various brood diseases with symptoms similar to that of foulbroods and sacbrood but with no predominant pathogen, AFB-like symptoms, spotty brood pattern, increased supersedure of queens, bees crawling on the ground, and a low adult bee population.

Bee Quick = A chemical, that smells like benzaldehyde that is used to drive bees from supers.

Bee space = A space between $1/4$ and $3/8$ inch which permits free passage for a bee but too small to encourage comb building, and too large to induce propolizing.

Bee suit = A pair of white coveralls made for beekeepers to protect them from stings and keep their clothes clean. Most come with zip-on veils.

Bee tree = A hollow tree occupied by a colony of bees.

Bee vac aka Bee vacuum = A vacuum used to suck up bees when doing a cutout or removal. Usually converted from a shop vac. It needs careful adjustment to not kill the bees.

Bee veil = Netting or screen for protecting the beekeeper's head and neck from stings.

Bee venom = The poison secreted by special glands attached to the stinger of the bee which is injected into the victim of a sting.

Beehive = A box usually with movable frames, used for housing a colony of bees.

Beelining = Finding feral bees by establishing the line which the bees fly back to their home. This can also include marking and timing the bees to get the distance and triangulating the location by releasing the bees from various places.

Beek = Beekeeper

Beekeeper = One who keeps bees. An Apiarist.

Beeswax = A substance that is secreted by bees by special glands on the underside of the abdomen, deposited as thin scales, and used after mastication and mixture with the secretion of the salivary glands for constructing the honeycomb. The melting point of beeswax is 144 to 147 °F.

Better Queens method = A graftless queen rearing method similar to Isaac Hopkins' actual queen rearing method (as opposed to the "Hopkins Method"). Sort of the Alley Method but with new comb instead of old.

Betterbee = A beekeeping supply company out of New York. They have many things no one else does. They also have eight frame equipment.

Benzaldehyde = A colorless nontoxic liquid aldehyde C_6H_5CHO that has an odor like that of bitter almond oil, that occurs in many essential oils and is sometimes used to drive bees out of honey supers. Also the flavor added to Maraschino cherries. What Bee Quick smells like.

Black scale = Refers to dried pupa, which died of American foulbrood.

Boardman feeder = A feeder that goes in the entrance and hold an inverted quart mason jar. They are notorious for causing robbing.

Bottling tank = A tank holding several gallons of honey, equipped with a honey gate to fill honey jars.

Bottom bar = The horizontal piece of the frame that is on the bottom of the frame.

Bottom board = The floor of a bee hive.

Bottom board feeder = This is picture of the bottom board feeder that Jay Smith came up with. It's simply a dam made with a $^3/_4$" by $^3/_4$" block of wood put an inch or so back from where the front of the hive would be (18" or so forward of the very back). The box is slid forward enough to make a gap at the back. The syrup is poured in the back. A small board can be used to block the opening in the back. The bees can still get out the front by simply coming down forward of the dam. The picture is from the perspective of standing behind the hive looking toward the front. The edges of the dam have been enhanced and labels put on to try to make more sense. This version doesn't work on a weak hive as the syrup is too close to the entrance. It drowns as many bees as the frame feeders.

Bottom supering = The act of placing honey supers under all the existing supers, directly on top of the brood box. The theory is the bees will work it better when it's directly above the brood chamber; as opposed to *top* supering which would be just putting the supers on top of the existing supers.

Box Jig = Jig for nailing boxes. (for more pictures see chapter by that name in Volume 3)

Brace comb = A bit of comb built between two combs to fasten them together, between a comb and adjacent wood, or between two wooden parts such as top bars.

Braula coeca = A wingless fly commonly known as the bee louse.

Breeder hive = The hive from which eggs or larvae are taken for queen rearing. In other words the donor hive.

Bricks = Used to keep the lids from blowing off in the wind and often used in particular configurations as visual clues as to the state of a hive.

Brood = Immature bees not yet emerged from their cells; in other words, egg, larvae or pupae.

Brood chamber = The part of the hive in which the brood is reared; may include one or more hive bodies and the combs within. Sometimes used to refer to a deep box as these are commonly used for brood.

Brood nest = The part of the hive interior in which brood is reared; usually the two bottom boxes.

Bt = Bacillus thuringiensis. A naturally occurring bacteria that is sprayed on empty comb to kill wax moths. Also sold to control larvae of other specific insects.

Buckfast = A strain of bees developed by Brother Adam at Buckfast Abbey in England, bred for disease resistance, disinclination to swarm, hardiness, comb building and good temper.

Burr comb = Small pieces of comb outside of the normal space in the frame where comb usually is. Brace comb would fall into this category.

C

Candy plug = A fondant type candy placed in one end of a queen cage to delay her release.

Capped brood = Immature bees whose cells have been sealed over with papery caps.

Capping melter = Melter used to liquefy the wax from cappings as they are removed from honeycombs.

Cappings = The thin wax covering over honey; once cut off of extracting frames.

Capping scratcher = A fork-like device used to remove wax cappings covering honey, so it can be extracted. Usually used on low areas that get missed by the uncapping knife.

Carniolan bees = Apis mellifera carnica. These are darker brown to black. They fly in slightly cooler weather and in theory are better in northern climates. They are reputed by some to be less productive than Italians, but I have not had that experience. The ones I have had were very productive and very frugal for the winter. They winter in small clusters and shut down brood rearing when there are dearths.

Castes = The three types of bees that comprise the adult population of a honey bee colony: workers, drones, and queen

Carts = Used for wheeling boxes or hives around.

Caucasian bees = Apis mellifera caucasica. They are silver gray to dark brown. They do propolis excessively. It is a sticky propolis rather than a hard propolis. They often coat everything with this sticky kind of proplolis, like fly paper. They build up a little slower in the spring than the Italians. They are reputed to be more gentle than the Italians. Less prone to robbing. In theory they are less productive than Italians. I think on the average they are about the same productivity as the Italians, but since they rob less you get less of the really booming hives that have robbed out all their neighbors.

Cell = The hexagonal compartment of a honeycomb.

Cell bar = A wooden strip on which queen cups are suspended for rearing queen bees.

Cell cup = Base of an artificial queen cell, made of beeswax or plastic and used for rearing queen bees or an empty beginning of a queen cell that the bees often build for no reason.

Cell finisher = A hive used to finish queen cells i.e. take them from capped to just before emergence. Sometimes queenright, sometimes queenless.

Cell starter = A hive used to start queen cells i.e. take them from just grafted to capped. Sometimes a "swarm box" or sometimes just a queenless hive.

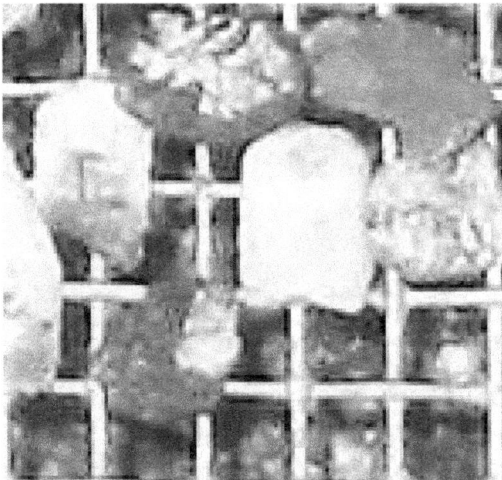

Chalkbrood = This is caused by a fungus Ascosphaera apis. It arrived in the US in 1968. If you find white pellets in front of the hive that kind of look like small corn kernels, you probably have chalkbrood. Putting the hive in full sun and adding more ventilation usually clears this up. Honey instead of syrup may contribute to clearing this up, since sugar syrup is much more alkaline (higher pH) than honey.

Checkerboarding (aka Nectar Management) = A method of swarm control and hive management pioneered by the late Walt Wright, that involves putting alternating frames of capped honey and empty drawn comb above the brood nest in late winter.

Chest hive = a hive that is laid out horizontally instead of vertically.

Chilled brood = Immature bees that have died from exposure to cold; commonly caused by mismanagement or sudden cold spells.

Chimney = When the bees fill only the center frames of honey supers.

Chinese grafting tool = Grafting tool made of plastic, horn and bamboo that has a retractable "tongue" that slides under the larvae and, when released, pushes it off of the "tongue". Popular because it is easier to operate than most grafting needles and it lifts up more royal jelly in the process. Quality varies and most recommend buying several and picking the ones you like out of those.

Chitin = Material which the exoskeleton of an insect is made of.

Chronic Paralysis Virus aka CPV = Symptoms: bees trembling, unable to fly, with K-wings and distended abdomens. One variety called the hairless black syndrome, is recognized by hairless, black shiny bees crawling at the hive entrance.

Chunk honey = Cut comb honey packed into jars then filled with liquid honey.

Clarifying = Removing visible foreign material from honey or wax to increase its purity.

Clipping = The practice of taking part of one or both wings off of a queen both for discouraging or slowing swarming and for identification of the queen.

Cloake Board AKA FWOF (Floor without a floor) = A device to divide a colony into a queenless cell starter and reunite it as a queenright cell finisher without having to open the hive.

Cloake board

Cluster = The thickest part of the bees on a warm day, usually the core of the brood nest. On a day below 50º F the only location where the bees are. It is used to refer both to the location and to the bees in that location.

Cocoon = A thin silk covering secreted by larval honey bees in their cells in preparation for pupation.

Coffin hive = a hive that is laid out horizontally instead of vertically.

Colony = The superorganism made up of worker bees, drones, queen, and developing brood living together as a family unit.

Colony Collapse Disorder = A recently named problem where most of the bees in most of the hives in an apiary disappear leaving a queen, healthy brood and only a few bees in the hive with plenty of stores.

Comb = The wax structures in a colony in which eggs are laid, and honey and pollen are stored. Shaped like hexagons.

Comb foundation = A commercially made structure consisting of thin sheets of beeswax with the cell bases of a particular cell size embossed on both sides to induce the bees to build a that size of cell.

Comb Honey = Honey in the wax combs, either cut from larger combs or produced and sold as a separate unit, such as a wooden section $4^1/_2''$ square, or a plastic round ring.

Conical escape = A cone-shaped bee escape, which permits bees, a one-way exit; used in a special escape board to free honey supers of bees.

Cordovan bees = A subset of the Italians. In theory you could have a Cordovan in any breed, since it's technically just a color, but the ones for sale in North American that I've seen are all Italians. They are slightly more gentle, slightly more likely to rob and quite striking to look at. They have no black on them and look very yellow at first sight. Looking closely you see that where the Italians normally have black legs and head, they have a purplish brown legs and head.

Creamed honey = Honey that has undergone controlled granulation to produce a finely textured candied or crystallized honey which spreads easily at room temperature. This usually involves adding fine "seed" crystals and keeping at 57º F (14º C).

Crimp-wired foundation = Comb foundation into which crimp wire is embedded vertically during foundation manufacture.

Crimper = A device used to put a ripple in the frame wire to both make it tight and to distribute stress better and give more surface to bind it to the wax.

Cupralarva = A particular brand of graftless queen rearing system.

Cut-comb Honey = Comb honey cut into various sizes, the edges drained, and the pieces wrapped or packed individually

Cut-out = Removing a colony of bees from somewhere that they don't have movable comb by cutting out the combs and tying them into frames.

D

Dadant = A beekeeping supply company out of Illinois. Founded by C.P. Dadant who was a pioneer in the modern beekeeping era and invented, among other things, the Jumbo and the square Dadant box. ($19^7/_8$" by $19^7/_8$" by $11^5/_8$"), published and wrote for the American Bee Journal and translated *Huber's Observations on Bees* from French to English and published many books including but not limited to the later versions of *The Hive and the Honey Bee*.

Dadant deep = A box designed by C.P. Dadant that is $11\ ^5/_8$" deep and the frame is $11^1/_4$" deep. Sometimes called Jumbo or Extra Deep.

Dearth = A period of time when there is no available forage for bees, due to weather conditions (rain, drought) or time of year.

Decoy hive aka Bait hive aka Swarm trap = A hive placed to attract stray swarms.

Deep = In Langstroth terms, a box that is $9^5/_8''$ deep and the frame is $9^1/_4''$ deep. Sometimes called a Langstroth Deep.

Deformed Wing Virus = A virus spread by the Varroa mite that causes crumpled looking wings on fuzzy newly emerged bees.

Demaree = The method of swarm control that separates the queen from most of the brood within the same hive and causes them to raise another queen with the goal of a two queen hive, increased production and reduced swarming.

Depth = The vertical measurement of a box or frame.

Dequeen = To remove a queen from a colony. Usually done before requeening, or as a help for brood diseases or pests.

Detritus = Wax scales and debris that sometimes build up at the bottom of a natural colony.

Dextrose = Also known as glucose, it is a simple sugar (or monosaccharide) and is one of the two main sugars found in honey; forms most of the solid phase in granulated honey.

Diastase = A starch digesting enzyme in honey adversely affected by heat; used in some countries to test quality and heating history of stored honey.

Diploid = Possessing pairs of genes, as workers and queens do, as opposed to haploid, possessing single genes as drones do.

Disease resistance = The ability of an organism to avoid a particular disease; primarily due to genetic immunity or avoidance behavior.

Dividing = Separating a colony to form two or more colonies. AKA a split

Division = Separating a colony to form two or more colonies.

Division board = A wooden or plastic piece like a frame but tight all the way around used to divide one box into more compartments for nucs.

Division board feeder or Frame feeder = A wooden or plastic compartment which is hung in a hive like a frame and contains sugar syrup to feed bees. The original designation (Division) was because it was *used* to make a division between two halves of a box to divide it into nucs, usually for queen rearing or making increase (splits). Most of them have a beespace around them now and cannot be used to make a division.

Domestic = Bees that live in a manmade hive. Since all bees are pretty much wild this is a relative term.

Doolittle method = A method of queen rearing that involves grafting young larvae into queen cups. First discovered by Nichel Jacob in 1568, then written about by Schirach in 1767 and then Huber in 1794 and finally popularized by G.M. Doolittle in his book *Scientific Queen Rearing* in 1846.

Double screen = A wooden frame, $^1/_2$ to $^3/_4$″ thick, with two layers of wire screen to separate two colonies within the same hive, one above the other. Often an entrance is cut on the upper side and placed to the rear of the hive for the upper colony and sometimes other openings are incorporated which would then be a Snelgrove board.

Double story or Double deeps = Referring to a beehive wintering in two deep boxes.

Double wide = A box that is twice as wide as a ten frame box. $32^1/_2$" wide.

Drawn combs = Full depth comb ready for brood or nectar with the cell walls drawn out by the bees, completing the comb as opposed to foundation that has not been worked by the bees and has no cell walls yet.

Drifting = The movement of bees that have lost their location and enter hives other than their own home. This happens often when hives are placed in long straight rows where returning foragers from the center hives tend to drift to the row ends or when making splits and the field bees drift back to the original hive.

"The percentage of foragers originating from different colonies within the apiary ranged from 32 to 63 percent"—from a paper, published in 1991 by Walter Boylan-Pett and Roger Hoopingarner in Acta Horticulturae 288, 6th Pollination Symposium (see Jan 2010 edition of Bee Culture, 36)

Drone = The male honey bee which comes from an unfertilized egg (and is therefore haploid) laid by a queen or less commonly, a laying worker.

Drone comb = Comb that is made up of cells larger than worker brood, usually in the range of 5.9 to 7.0mm in which drones are reared and honey and pollen are stored.

Drone brood = Brood, which matures into drones, reared in cells larger than worker brood. It is noticeably larger than worker brood and the cappings are distinctly dome shaped.

Drone Congregation Area = A place that drones from many surrounding hives congregate and wait for a queen to come. In other words a mating area. Drones find them by following both pheromone trails and topographical features of the landscape such as tree rows.

Drone layers = A drone laying queen (one with no sperm left to fertilize eggs) or laying workers.

Drone laying queen = A queen that can lay only unfertilized eggs, due to age, improper or late mating, disease or injury.

Drone mother hive = The hive which is encouraged to raise a lot of drones to improve the drone side of mating queens. Based on the myth that you can make bees raise more drones. Taking drone comb from the ones you want to perpetuate and giving them to other colonies is the only real way to succeed at this as the mother colony will then raise more drones while the colonies receiving the drone comb will raise less of their own because they will be raising the ones from the drone mother.

Drumming = Tapping or thumping on the sides of a hive to make the bees ascend into another hive placed over it or to drive them out of a tree or house. This will not get all of them out, but will move a significant number.

Dorsal-Ventral Abdominal Vibrations dance = A dance used to recruit forages. Also used on queen cells about to emerge and possibly other times.

Dwindling = Any rapid decline in the population of the hive. The rapid dying off of old bees in the spring; sometimes called spring dwindling or disappear-ing disease.

Dysentery = A condition of adult bees character-ized by severe diarrhea (as evidenced by brown or yellow streaks on the front of the hive) and usually caused by long confinement (from either cold or bee-keeper manipulation), starvation, low-quality food, or Nosema infection.

E

Eight frame = Boxes that were made to take eight frames. Usually between $13^1/_2$" and 14" wide depending on the manufacturer. Typically $13^3/_4$" wide.

Eggs = The first phase in the bee life cycle, usu-ally laid by the queen, is the cylindrical egg $^1/_{16}$" (1.6 mm) long; it is enclosed with a flexible shell or chorion. It resembles a small grain of rice.

Eke = The term originated with skeps and it was "an enlargement" which is the equivalent of today's super. In current usage it usually refers to a shim that is either added to the top for feeding things like pollen patties or added under a shallow to make it into a deep. The term is used more frequently in Britain.

Electric embedder = A device that heats the foundation wire by running current through it for em-bedding of wires in foundation.

End bar = The piece of a frame that is on the ends of the frame i.e. the vertical pieces of the frame.

Entrance reducer = A wooden strip used to reg-ulate the size of the entrance.

Escape board = A board having one or more bee escapes in it used to remove bees from supers.

European Foulbrood = Caused by a bacteria. It used to be called Streptococcus pluton but has now been renamed Melissococcus pluton. European Foul Brood is a brood disease. With EFB the larvae turn brown and their trachea is even darker brown. Don't confuse this with larvae being fed dark honey. It's not just the food that is brown. Look for the trachea. When it's worse, the brood will be dead and maybe black and maybe sunk cappings, but usually the brood dies before they are capped. The cappings in the brood nest will be scattered, not solid, because they have been removing the dead larvae. To differentiate this from AFB use a stick and poke a diseased larvae and pull it out. The AFB will "string" two or three inches.

Ether wash = Putting a cupful of bees in a jar with a spray of starter fluid to kill the bees and mites so you can count the Varroa mites. A sugar roll is a non-lethal and much less flammable method of doing the same.

European Honey Bees = Bees from Europe as opposed to bees originating in Africa or other parts of the world or bees crossbred with those from Africa.

Eyelets = Optional small metal piece fitting into the wire-holes of a frame's end bar; used to keep the reinforcing wires from cutting into the wood. Many people use a staple across where it would split the wood instead.

Extra shallow = A box that is $4^{11}/_{16}$ or $4^3/_4''$ deep. Usually used for cut comb. Sometimes modified for sections.

Extracted honey = Honey removed from combs usually by means of a centrifugal force (an extractor) in order to leave the combs intact but with hobbyists often from crushing the comb and straining it (see Crush and Strain).

Ezi Queen = A particular brand of graftless queen rearing system.

F

Frame feeder or division board feeder = A wooden or plastic compartment which is hung in a hive like a frame and contains sugar syrup to feed bees. The original designation (Division) was because it was *used* to make a division between two halves of a box to divide it into nucs, usually for queen rearing or making increase (splits). Most of them have a beespace around them now and cannot be used to make a division.

Feeders = Any device used to feed bees.

Fermenting honey = Honey which contains too much water (greater than 20%) in which yeast has grown and caused some of it to turn into carbon dioxide, water and alcohol.

Feral (queen or bees) = Since all North American bees are considered to have come from domestic stock, what most people call "wild" bees are really "feral" bees. Some use the term for survivor bees that were captured and used to raise queens meaning they *were* feral as opposed to *are* feral.

Fertile queen = An inseminated queen.

Fertilized = Usually refers to eggs laid by a queen bee, they are fertilized with sperm stored in the queen's spermatheca, in the process of being laid. These develop into workers or queens.

Festooning = The activity of young bees, engorged with honey, hanging on to each other usually to secrete beeswax but also in bearding and swarming..

Field bees = Worker bees which are usually 21 or more days old and work outside to collect nectar, pollen, water and propolis; also called foragers.

Flash heater = A device for heating honey very rapidly to prevent it from being damaged by sustained periods of high temperature

Flight path = Usually refers to the direction bees fly leaving their colony; if obstructed, may cause bees to accidentally collide with the person obstructing and eventually become aggravated.

Floor Without a Floor AKA FWOF AKA Cloake Board = A device to divide a colony into a queenless cell starter and reunite it as a queenright cell finisher without having to open the hive.

Follower board = A thin board used in place of a frame usually when there are fewer than the normal number of frames in a hive. This is usually referring to one that has a beespace around it and is used to make the frames easier to remove without rolling and to cut down on condensation on the walls. Sometimes it's used to refer to a board that is bee tight and used to divide a box into two colonies. When designed and used in this manner it should be called a division board.

Food chamber = A hive body filled with honey for winter stores. Typically a third deep used in unlimited brood nest management.

Forage = Natural food source of bees (nectar and pollen) from wild and cultivated flowers. Or the act of gathering that food.

Foragers = Worker bees which are usually 21 or more days old and work outside to collect nectar, pollen, water and propolis; also called field bees.

Foundation = Thin sheets of beeswax embossed or stamped with the base of a worker (or rarely drone) cells on which bees will construct a complete comb (called drawn comb); also referred to as comb foundation, it comes wired or unwired and also in plastic as well as one piece foundations and frames as well as different thicknesses (thin surplus, surplus, medium) and different cell sizes (brood =5.4mm, small cell = 4.9mm, drone=6.6mm).

Foundationless = A frame with some kind of comb guide that is used without foundation.

Frame = A rectangular structure of wood designed to hold honeycomb, consisting of a top bar, two end bars, and a bottom bar; usually spaced a bee-space apart in the super.

Frame feeder = Sometimes called a "division board feeder". It takes the place of one or more frames. Less bees drown if you put floats in.

Fructose = Fruit sugar, also called levulose (left handed sugar), a monosaccharide commonly found in honey that is slow to granulate

Fumagilin-B = Bicyclohexyl-ammonium fumagillin, whose trade name was Fumidil-B (Abbot Labs) but now seems to be called Fumagilin-B, is a whitish soluble antibiotic powder discovered in 1952; some beekeepers mix this with sugar syrup and feed it to bees to control Nosema disease. Fumagilin is more soluble than Fumidil. Its use in beekeeping is outlawed in the European Union because it is a suspected teratogen (causes birth defects). Fumagilin can block blood vessel formation by binding to an enzyme called methionine aminopeptidase. Targeted gene disruption of methionine aminopeptidase 2 results in an embryonic gastrulation defect and endothelial cell growth arrest. It is made from the fungus that causes stonebrood, Aspergillus fumigatus. Formula: (2E, 4E, 6E, 8E)– 10- {[(3S, 4S, 5S, 6R)- 5– methoxy- 4- [2–methyl– 3- (3–methylbut– 2-enyl) oxiran– 2-yl]- 1- oxaspiro[2.5] octan- 6- yl] oxy}- 10- oxo- deca- 2, 4, 6, 8- tetraenoic acid

Fumidil-B = The old trade name for Fumagilin, see above entry.

Fume board = A device used to hold a set amount of a volatile chemical (A bee repellent like Bee Go or Honey Robber or Bee Quick) to drive bees from supers.

G

Gloves = Leather, cloth or rubber gloves worn while inspecting bees.

Glucose = Also known as dextrose, it is a simple sugar (or monosaccharide) and is one of the two main sugars found in honey; forms most of the solid phase in granulated honey.

Grafting = Removing a worker larva from its cell and placing it in an artificial queen cup in order to have it reared into a queen.

Grafting tool = A needle or probe used for transferring larvae in grafting of queen cells.

Granulate = The process by which honey, a super-saturated solution (more solids than liquid) will become solid or crystallize; speed of granulation depends of the kinds of sugars in the honey, the crystal seeds (such as pollen or sugar crystals) and the temperature. Optimum temperature for granulation is 57º F (14º C).

Guard bees = Worker bees about three weeks old, which have their maximum amount of alarm pheromone and venom; they challenge all incoming bees and other intruders.

Gum = A hollow log beehive, sometimes called a log-gum, made by cutting out that portion of a tree containing bees and moving it to the apiary, or by cutting a hollow portion of a log, putting a board on for a lid and hiving a swarm in it. Since it contains no moveable combs, and since each individual state in the US has laws that require movable combs, it is therefore illegal in the US.

H

Hair clip queen catcher = A device used to catch a queen that resembles a hair clip. Available from most beekeeping supply houses.

Haploid = Possessing a single set of genes, as drones do, as opposed to pairs of genes as workers and queens have.

Hemolymph = The scientific name for insect "blood."

Hive = A home for a colony of bees.

Hive body = A wooden box containing frames. Usually referring to the size of box being used for brood.

Hive stand = A structure serving as a base support for a beehive; it helps in extending the life of the bottom board by keeping it off damp ground. Hive stands may be built from treated lumber, cedar, bricks, concrete blocks etc.

Hive staples = Large C-shaped metal nails, hammered into the wooden hive parts to secure bottom to supers, and supers to super before moving a colony.

Hive tool = A flat metal device used to pry boxes and frames apart, typically with a curved scraping surface or a lifting hook at one end and a flat blade at the other.

Hoffman frame = Frames that have the end bars wider than the top bars to provide the proper spacing when frames are placed in the hive. In other words, self-spacing frames. In other words, standard frames.

Honey = A sweet viscous material produced by bees from the nectar of flowers, composed largely of a mixture of dextrose and levulose dissolved in about 19 to 17 percent water; contains small amounts of sucrose, mineral matter, vitamins, proteins, and enzymes.

Honey bound = A condition where the brood nest of a hive is being backfilled with honey. This is a normal condition that is used by the workers to shut down the queen's brood production. It usually happens just before swarming and in the fall to prepare for winter.

Honeydew = An excreted material from insects in the order Homoptera (aphids) which feed on plant sap; since it contains almost 90% sugar, it is collected by bees and stored as honeydew honey.

Honey bee = The common name for Apis mellifera.

Honey Bee Healthy = A mixture of essential oils (lemon grass and peppermint) sold to boost the immune system of the bees.

Honey crop = The honey that was harvested.

Honey crop also called honey stomach or honey sac = An enlargement at the posterior of a bees' esophagus but lying in the front part of the abdomen, capable of expanding when full of liquid such as nectar or water. Used for transportation purposes for water, nectar and honey.

Honey extractor = A machine which removes honey from the cells of comb by centrifugal force. The two main types are tangential where the frames lie flat and are flipped to extract the other side, and radial where the frames are like spokes in a wheel and both sides are emptied at the same time.

Honey flow = A time when enough nectar-bearing plants are blooming such that bees can store a surplus of honey.

Honey gate = A faucet used for removing honey from tanks and other storage receptacles.

Honey house = A building used for activities such as honey extraction, packaging and storage.

Honey plants = Plants whose flower (or other parts) yields enough nectar to produce a surplus of honey; examples are asters, basswood, citrus, eucalyptus, goldenrod and tupelo.

Honey Super Cell = Fully drawn plastic comb in deep depth and 4.9mm cell size

Honey supers = Refers to boxes of frames used for honey production. From the Latin "super" for above as a designation for any box above the brood nest.

Hopkins method = A graftless method of queen rearing that involves putting a frame of young larvae horizontally above a brood nest.

Hopkins shim = A shim used to turn a frame flatways for queen rearing without grafting.

Horizontal hive = a hive that is laid out horizontally instead of vertically in order to eliminate lifting boxes.

Hornets and Yellow Jackets = Social insects belonging to the family Vespidae. Nest in paper or foliage material, with only an overwintering queen. Fairly aggressive, and carnivorous, but generally beneficial, they can be a nuisance to man. Hornets and Yellow Jackets are often confused with Wasps and honey bees. Wasps are related to hornets and yellow jackets, the most common of which are the paper wasps which nest in small exposed paper combs, suspended by a single support. Hornets, yellow jackets and wasps are easy to distinguish by their shiny hairless body, and aggressiveness. Yellow jackets, unfortunately, look like the bees in the cartoons and advertisements, bright yellow and black and shiny. Honey bees are generally fuzzy black, brown or tan, never bright yellow, and basically docile in nature.

Hot (temperament) = Bees that are overly defensive or outright aggressive.

Housel positioning theory = A theory proposed by Michael Housel that natural brood nests have a predictable orientation of the "Y" in the bottom of the cells.

Hydroxymethyl furfural = A naturally occurring compound in honey that rises over time and rises when honey is heated.

Hypopharyngeal gland = A gland located in the head of a worker bee that secretes "royal jelly". This rich blend of proteins and vitamins is fed to all bee larvae for the first three days of their lives and queens during their entire development.

I

Israeli Acute Paralysis Virus aka IAPV = The virus currently being blamed for CCD. First discovered in Israel where it was quite devastating to colonies.

Illinois = A box that is $6^5/_8$" in depth and the frames are $6^1/_4$" in depth. AKA Medium AKA Western AKA $^3/_4$ depth.

Imirie shim = A device credited to the late George Imirie that is a $^3/_4$" shim with an entrance built in. It allows you to add an entrance between any two pieces of equipment on the hive.

Increase = To add to the number of colonies, usually by dividing those on hand. See Split.

Infertile = Incapable of producing a fertilized egg, as a laying worker or drone laying queen. Unfertilized eggs develop into drones.

Inhibine = Antibacterial effect of honey caused by enzymes and an accumulation of hydrogen peroxide, a result of the chemistry of honey.

Inner cover = An insulating cover fitting on top of the top super but underneath the outer cover, typically with an oblong hole in the center. Used to be called a "quilt board". In the old days these were often made of cloth.

Instar = Stages of larval development. A honey bee goes through five instars. The best queens are grafted in the 1st (preferably) or 2nd instar and not later than that.

Instrumental insemination aka II or AI = The introduction of drone spermatozoa into the spermatheca of a virgin queen by means of special instruments

Invertase = An enzyme in honey, which splits the sucrose molecule (a disaccharide) into its two components dextrose and levulose (monosaccharides). This is produced by the bees and put into the nectar to convert it in the process of making honey.

Isomerase = A bacterial enzyme used to convert glucose in corn syrup into fructose, which is a sweeter sugar; called isomerose, is now used as a bee feed.

Italian bees = A common race of bees, Apis mellifera ligustica, with brown and yellow bands, from Italy; usually gentle and productive, but tend to rob and brood incessantly.

J

Jenter = A particular brand of graftless queen rearing system.

K

Kashmir Bee Virus = A widespread disease of bees, spread more quickly by Varroa, found everywhere there are bees.

Kenya Top Bar Hive = A top bar hive with sloped sides. The theory is that they will have less attachments on the sides because of the slope.

Kidneys = Bees don't actually have kidneys. They have malpighian tubules which are thin filamentous projects from the junction of the mid and hind gut of the bee that cleanse the hemolymph (blood) of nitrogenous cell wastes and deposit them as non-toxic uric acid crystals into the undigestible food wastes for elimination. They serve the same purpose in bees as kidneys do in higher animals.

L

Landing board = An extraneous construction that makes a small platform at the entrance of the hive for the bees to land on before entering the hive. Usually just a longer bottom board. Sometimes a sloped approach is added. Bees in nature have none. I call it a "mouse ramp" as the only actual purpose I see it provide is a place for mice to get into the hive more conveniently.

Lang = Short for Langstroth hive.

Langstroth, Rev. L.L. = A Philadelphia native and minister (1810-95), he lived for a time in Ohio where he continued his studies and writing of bees; recognized the importance of the bee space, resulting in the development of the most commonly used movable-frame hive.

Langstroth hive = The basic hive design of L.L. Langstroth. In modern terms any hive that takes frames that have a 19" top bar and fit into a box $19^7/_8$" long. Widths vary from five frame nucs to eight frame boxes to ten frame boxes and from Dadant deeps, Langstroth deeps, Mediums, Shallows and Extra Shallow. But all would still be Langstroths. This would distinguish them from WBC, Smith, National, DE etc.

Large Cell = Standard foundation size = 5.4mm cell size

Larva, open = The second developmental stage of a bee, starting the 4th day from when the egg is laid until it's capped on about the 9th or 10th day.

Larva, capped = The second developmental stage of a bee, ready to pupate or spin its cocoon (about the 10th day from the egg).

Laying workers = Worker bees which lay eggs in a colony caused by them being a few weeks without the pheromones from open brood; such eggs are infertile, since the workers cannot mate, and therefore become drones.

Leg baskets = Also called pollen baskets, a flattened depression surrounded by curved spines located on the outside of the tibiae of the bees' hind legs and adapted for carrying flower pollen and propolis.

Lemon Grass essential oil = Essential oil used for swarm lure which contains many of the constituents of Nasonov pheromone.

Levulose = Also called fructose (fruit sugar), a monosaccharide commonly found in honey that is slow to granulate.

Long hive = a hive that is laid out horizontally instead of vertically.

M

Malpighian tubules = Thin filamentous projections from the junction of the mid and hind gut of the bee that cleanse the hemolymph of nitrogenous cell wastes and deposit them as non-toxic uric acid crystals into the undigestible food wastes for elimination. They serve the same purpose as kidneys in higher animals.

Mandibles = The jaws of an insect; used by bees to form the honeycomb and scrape pollen, in fighting and picking up hive debris.

Marking = Painting a small dot of enamel on the back of the thorax of a queen to make her easier to identify and so you can tell her age and if she has been superseded.

Marking pen = An enamel pen used to mark queens. Available at local hardware stores as enamel pens. Also from beekeeping supply houses as Queen marking pens.

Marking Tube = A plastic tube commonly available from beekeeping supply houses that is used to safely confine a queen while you mark her.

Mating flight = The flight taken by a virgin queen while she mates in the air with several drones.

Mating nuc = A small nuc for the purpose of getting queens mated used in queen rearing.

Maxant = A beekeeping equipment manufacturer that makes uncappers, extractors, hive tools etc.

Medium = A box that is $6^5/_8$″ in depth and the frames are $6^1/_4$″ in depth. AKA Illinois AKA Western AKA $^3/_4$ depth.

Medium brood (foundation) = When used to refer to foundation, medium refers to the thickness of the wax *not* the depth of the frame. In this case it's medium thick and of worker sized cells.

Melissococcus pluton = New name given by taxonomists for the bacterium that causes European foulbrood. The old name was Streptococcus pluton.

Migratory beekeeping = The moving of colonies of bees from one locality to another during a single season to take advantage of two or more honey flows or for pollination.

Migratory cover = An outer cover used without an inner cover that does not telescope over the sides of the hive; used by commercial beekeepers who frequently move hives. This allows hives to be packed tightly against one another because the cover does not protrude over the sides.

Miller Bee Supply = A beekeeping supply company out of North Carolina *(www.millerbeesupply.com)*. Among other things, they have eight frame equipment.

Miller feeder = Top feeder popularized by C.C. Miller.

Miller Method = A graftless method of queen rearing that involves a ragged edge on some brood comb for the bees to build queen cells on.

Moisture content = In honey, the percentage of water should be no more than 18.6; any percentage higher than that will allow honey to ferment.

Mouse guard = A device to reduce the entrance to a hive so that mice cannot enter. Commonly #4 hardware cloth.

Movable combs = Combs that are built in a hive that allows them to be manipulated and inspected individually. Top bar hives have movable combs but not frames. Langstroth hives have movable combs *in* frames.

Movable frames = A frame constructed in such a way to preserve the bee space, so they can be easily removed; when in place, it remains unattached to its surroundings.

N

Nadiring = Adding boxes below the brood nest. This is a common practice with foundationless including Warre' hives.

Nasonov = A pheromone given off by a gland under the tip of the abdomen of workers that serves primarily as an orientation pheromone. It is essential to swarming behavior and nasonoving is set off by disturbance of the colony. It is a mixture of seven terpenoids, the majority of which is Geranial and Neral, which are a pair of isomers usually mixed and called citral. Lemongrass (Cymbopogon) essential oil is mostly these scents and is useful in bait hives and to get newly hived bees or swarms to stay in a hive.

Nasonoving = Bees who have their abdomens extended and are fanning the Nasonov pheromone. The smell is a mixture of lemon and geranium.

Natural cell = Cell size that bees have built on their own without foundation.

Natural comb = Comb that bees have built on their own without foundation.

Nectar = A liquid rich in sugars, manufactured by plants and secreted by nectary glands in or near flowers; the raw material for honey.

Nectar flow = A period of time when nectar is available.

Nectar Management aka Checkerboarding = a method of swarm control originated by the late Walt Wright where the stores above the brood chamber are alternated with drawn comb late in the winter. Reports from those using it are of massive harvests and no swarming.

New World Carniolans = A breeding program originated by Sue Cobey to find and breed bees from the US with Carniolan traits and other commercially useful traits.

Newspaper method = A technique to join together two colonies by providing a temporary newspaper barrier. Usually one sheet with a small slit. Usually you make sure both colonies can still fly and ventilate.

Nicot = A particular brand of graftless queen rearing system.

Nosema = Caused by a fungus (used to be classified as a protozoan) called Nosema apis. Nosema is present all the times and is really an opportunistic disease. The common chemical solution (which I don't use) was Fumidil which has been recently renamed Fumagilin-B. In my opinion the best prevention is to make sure your hive is healthy and not stressed and feed honey.

Nuc, nuclei, nucleus = A small colony of bees often used in queen rearing or the box in which the small colony of bees resides. The term refers to the fact that the essentials, bees, brood, food, a queen or the means to make one, are there for it to grow into a colony, but it is not a full sized colony.

Nurse bees = Young bees, usually three to ten days old, which feed and take care of developing brood.

O

Observation Hive = A hive made largely of glass or clear plastic to permit observation of bees at work

Open-air Nest = A colony that has built its nest in the open limbs of a tree rather than in the hollow of a tree or a hive.

Open Mesh Floor = British version of "screened bottom board".

Outer cover = The last cover that fits over a hive to protect it from rain; the two most common kinds are telescoping and migratory covers.

Outyard = Also called out apiary, it is an apiary kept at some distance from the home or main apiary of a beekeeper.

Ovary = The egg producing part of a plant or animal.

Ovule = An immature female germ cell, which develops into a seed.

Ovariole = Any of several tubules that compose an insect ovary.

Oxytetracycline aka Oxytet = An antibiotic sold under the trade name Terramycin; used to control American and European foulbrood diseases.

P

Package bees = A quantity of adult bees (2 to 5 pounds), with or without a queen, contained in a screened shipping cage.

Parasitic Mite Syndrome aka Bee Parasitic Mite Syndrome = A set of symptoms that are caused by a major infestation of Varroa mites. Symptoms include the presence of Varroa mites, the presence of various brood diseases with symptoms similar to that of foulbroods and sacbrood but with no predominant pathogen, AFB-like symptoms, spotty brood pattern, increased supersedure of queens, bees crawling on the ground, and a low adult bee population.

Parasitic Mites = Varroa and tracheal mites are the mites with economic issues for bees. There are several others that are not known to cause any problems.

Paralysis aka APV aka Acute Paralysis Virus = A viral disease of adult bees which affects their ability to use legs or wings normally.

Parthenogenesis = The development of young from unfertilized eggs laid by virgin females (queen or worker); in bees, such eggs develop into drones.

Para Dichloro Benzene (aka PDB aka Paramoth) = Wax moth treatment for stored combs. A known carcinogen.

PermaComb = Fully drawn plastic comb in medium depth and about 5.0mm equivalent cell size after allowing for cell wall thickness and taper of the cell..

PF100 (deep) and PF120 (medium) = A small cell one piece plastic frame available from Mann Lake. Measures 4.95mm cell size. Users report excellent acceptance and perfectly drawn cells.

Phoretic = In the context of Varroa mites it refers to the state where they are on the adult bees instead of in the cell either developing or reproducing.

Piping = A series of sounds made by a queen, frequently before she emerges from her cell. When the queen is still in the cell it sounds sort of like a quack quack quack. When the queen has emerged it sounds more like zoot zoot zoot.

Play flights aka orientation flights = Short flights taken in front and in the vicinity of the hive by young bees to acquaint them with hive location; sometimes mistaken for robbing or swarming preparations.

Pollen = The dust-like male reproductive cells (gametophytes) of flowers, formed in the anthers, and important as a protein source for bees; fermented pollen (bee bread) is essential for bees to rear brood.

Pollen basket = An anatomical structure on the bees legs where pollen and propolis is carried.

Pollen bound = A condition where the brood nest of a hive is being filled with pollen so that there is nowhere for the queen to lay.

Pollen pellets or cakes = The pollen packed in the pollen baskets of bees and transported back to the colony made by rolling in the pollen, brushing it off and mixing it with nectar and packing it into the pollen baskets.

Pollen substitute = A food material which is used to substitute wholly for pollen in the bees' diet; usually contains all or part of soy flour, brewers' yeast, wheast, powdered sugar, or other ingredients. Research has shown that bees raised on substitute are shorter lived than bees raised on real pollen.

Pollen supplement = A mixture of pollen and pollen substitutes used to stimulate brood rearing in periods of pollen shortage

Pollen trap = A device for collecting the pollen pellets from the hind legs of worker bees; usually forces the bees to squeeze through a screen mesh, usually #5 hardware cloth, which scrapes off the pellets which fall through #7 hardware cloth into a drawer with a screened bottom so the pollen won't mold.

Porter bee escape = Introduced in 1891, the escape is a device that allows the bees a one-way exit between two thin and pliable metal bars that yield to the bees' push.

Prime swarm = The first swarm to leave the parent colony, usually with the old queen.

Proboscis = The mouthparts of the bee that form the sucking tube or tongue

Propolis = Plant resins collected, mixed with enzymes from bee saliva and used to fill in small spaces inside the hive and to coat and sterilize everything in the hive. It has antimicrobial properties. It is typically made from the waxy substance from the buds of the poplar family.

Propolize = To fill with propolis, or bee glue.

Pupa = The third stage in the development of the bee during which it is inactive and sealed in its cocoon.

Push In Cage = Cage made of #8 hardware cloth used to introduce or confine queens to a small section of comb. Usually used over some emerging brood.

Q

Queen = A fully developed female bee responsible for all the egg laying of a colony.

Queen Bank = Putting multiple caged queens in a nuc or hive.

Queen cage = A special cage in which queens are shipped and/or introduced to a colony, usually with 4 to 7 young workers called attendants, and usually a candy plug.

Queen cage candy = Candy made by kneading powdered sugar with invert sugar syrup until it forms a stiff dough; used as food in queen cages.

Queen cell = A special elongated cell resembling a peanut shell in which the queen is reared; usually over an inch in length, it hangs vertically from the comb.

Queen clipping = Removing a portion of one or both wings of a queen to prevent her from flying or to better identify when she has been replaced.

Queen cup = A cup-shaped cell hanging vertically from the comb, but containing no egg; also made artificially of wax or plastic to raise queens

Queen excluder = A device made of wire, wood or zinc (or any combination thereof) having openings of .163 to .164 inch, which permits workers to pass but excludes queens and drones; used to confine the queen to a specific part of the hive, usually the brood nest.

Queenright = A colony that contains a queen capable of laying fertile eggs and making appropriate pheromones that satisfy the workers of the hive that all is well.

Queen Mandibular Pheromone aka Queen substance aka QMP = A pheromone produced by the queen and fed to her attendants who share it with the rest of the colony that gives the colony the sense of being queenright. Chemically QMP is very diverse with at least 17 major components and other minor ones. 5 of these compounds are: 9-ox-2-decenoic acid (9ODA) + cis & trans 9 hydroxydec-2-enoic acid (9HDA) + methyl-p-hydroxybenzoate (HOB) and 4-hydroxy-3-methoxyphenylethanol (HVA). Newly emerged queens produce very little of this. QMP is responsible for inhibition of rearing replacement queens, attraction of drones for mating, stabilizing and organizing a swarm around the queen, attracting a retinue of attendants, stimulating foraging and brood rearing, and the general moral of the colony. Lack of it also seems to attract robber bees.

Queen muff = A screen wire tube that resembles a "muff" to keep your hands warm in shape but is used to keep queens from escaping when marking them or releasing attendants.

R

Rabbet = In wood working a groove cut into wood. The frame rests in a Langstroth hive are rabbets and the corners are sometimes done as rabbets and sometimes as finger or box joints.

Races of Bees = In taxonomy this is actually a variety but in beekeeping it is typically called a "race". All of these are Apis mellifera. The most common currently In the US are Italians (ligustica), Carniolans (carnica) and Caucasians (caucasica). Russians would be either carpatica, acervorum, carnica or caucasica depending on who you are talking to.

Radial extractor = A centrifugal force machine to throw out honey but leave the combs intact; the frames are placed like spokes of a wheel, top bars towards the wall, to take advantage of the upward slope of the cells.

Rauchboy = A particular brand of smoker that has an inner chamber to provide more consistent oxygen to the fire.

Raw honey = Honey that has not been finely filtered or heated.

Regression = As applied to cell size, large bees, from large cells, cannot build natural sized cells. They build something in between. Most will build 5.1 mm worker brood cells. Regression is getting large bees back to smaller bees so they can and will build smaller cells.

Reorientation = When the bees take note of their surroundings and landmarks to make sure they remember the location of the colony. A variety of things set this off. Young bees will orient (not reorient but it's the same behavior) when they first emerge from the hive.

Requeen = To replace an existing queen by removing her and introducing a new queen.

Rendering wax = The process of melting combs and cappings and removing refuse from the wax.

Retinue = Worker bees that are attending the queen.

Reversing aka Switching = The act of exchanging places of different hive bodies of the same colony; usually for the purpose of nest expansion. The super full of brood and the queen is placed below an empty super to allow the queen extra laying space.

Robber screen = A screen used to foil robbers but let the local residents into the hive.

Robbing = The act of bees stealing honey/nectar from the other colonies; also applied to bees cleaning out wet supers or cappings left uncovered by beekeepers and sometimes used to describe the beekeeper removing honey from the hive.

Ropy = A quality of forming an elastic rope when drawn out with a stick. Used on capped brood as a diagnostic test for American foulbrood.

Round sections = Sections of comb honey in plastic round rings instead of square wooden boxes, usually Ross Rounds.

Rolling = A term to describe what happens when a frame is too tight or pulled out too quickly and bees get pushed against the comb next to it and "rolled".

This makes bees very angry and is sometimes the cause of a queen being killed.

Royal jelly = A highly nutritious, milky white secretion of the hypopharyngeal gland of nurse bees; used to feed the queen and young larvae.

Russian bees = Apis mellifera acervorum or carpatica or caucasica or carnica. They came from the Primorsky region of Russia and were used for breeding mite resistance because they were already surviving the mites. They were brought to the USA by the USDA in June of 1997, studied on an island in Louisiana and then field testing in other states in 1999. They went on sale to the general public in 2000.

S

Sac Brood Virus = Symptoms are the spotty brood patterns as other brood diseases but the larvae are in a sack with their heads raised.

Sclerite = Same as Tergite. An overlapping plate on the dorsal side of a arthropod that allows it to flex.

Screened Bottom Board = A bottom board with screen for the bottom to allow ventilation and to allow Varroa mites to fall through. AKA Open Mesh Floor.

Scout bees = Worker bees searching for a new source of pollen, nectar, propolis, water, or a new home for a swarm of bees.

Scutum = Shield shaped portion of the back of the thorax of some insects including Apis mellifera (honey bees). Usually divided into three areas: the anterior prescutum, the scutum, and the smaller posterior scutellum.

Sections = Small wooden (or plastic) boxes used to produce comb honey.

Self-spacing frames aka Hoffman frames = Frames constructed so that everything but the end bar (which is the spacer) is a bee space apart when pushed together in a hive body.

Settling tank = A large capacity container used to settle extracted honey; air bubbles and debris will float to the top, clarifying the honey.

Shallow = A box that is $5^{11}/_{16}$ or $5^3/_4$" deep with frames that are $5^1/_2$" deep.

Shaken swarm = An artificial swarm made by shaking bees off of combs into a screened box and then putting a caged queen in until they accept her. One method for making a divide. Also the method used to make packages of bees.

Skep = A beehive without movable combs, usually made of twisted straw in the form of a basket; its use is illegal in each state in the U.S as the combs are not inspectable.

Slatted rack = A wooden rack that fits between the bottom board and hive body. Bees make better use of the lower brood chamber with increased brood rearing, less comb gnawing, and less congestion at the front entrance. Popularized by C.C. Miller and Carl Killion.

Slumgum = The refuse from melted combs and cappings after the wax has been rendered or removed; usually contains cocoons, pollen, bee bodies and dirt.

Small Cell = 4.9mm cell size. Used by some beekeepers to control Varroa mites.

Small Hive Beetle = A pest recently imported to North America, whose larvae will destroy comb and ferment honey.

Smith method = A method of queen rearing popularized by Jay Smith, that uses a swarm box as a cell starter and grafting larvae into queen cups.

Smoker = A metal container with attached bellows which burns various fuels to generate smoke; used to interfere with the ability to smell alarm pheromone and therefore control aggressive behavior of bees during colony inspections.

Solar wax melter = A glass-covered box used to melt wax from combs and cappings using the heat of the sun.

Sperm cells = The male reproductive cells (gametes) which fertilize eggs; also called spermatozoa.

Spermatheca = A small sac connected with the oviduct of the queen bee in, which is stored, the spermatozoa received by the queen when mating with drones.

Spiracles = Openings into the respiratory system on a bee that can be closed at will. These are on the sides of the bee. They are considerably smaller than the Trachea they protect. The first thoracic spiracle is the one that is infiltrated by the tracheal mites as it is the largest. When closed the spiracles are air tight.

Split = To divide a colony for the purpose of increasing the number of hives.

Spur embedder = A device used for mechanically embedding wires into foundation by employing hand pressure as opposed to using electricity to melt the wires into the wax.

Starter hive aka a Swarm box = A box of shaken bees used to start queen cells.

Sting = An organ belonging exclusively to female insects developed from egg laying mechanisms, used to defend the colony; modified into a piercing shaft through which venom is injected.

Streptococcus pluton = Deprecated (old) name for the bacterium that causes European foulbrood. The new name is Melissococcus pluton.

Sucrose = A polysaccharide. The principal sugar found in nectar. Honey bees break this into Dextrose and Fructose with enzymes.

Sugar syrup = Feed for bees, containing sucrose or table (cane or beet) sugar and hot water in various ratios; usually 1:1 in the spring and 2:1 in the fall.

Sugar roll test = A test for Varroa mites that involves rolling a cupful of bees in powdered sugar and counting the number of mites dislodged.

Super = A box with frames in which bees store honey; usually placed above the brood nest. From the Latin *super* meaning "above".

Supering = The act of placing honey supers on a colony in expectation of a honey flow.

Supersedure = Rearing a new queen to replace the mother queen in the same hive.

Suppressed Mite Reproduction aka SMR = Queens from a breeding program by Dr. John Harbo that have less Varroa problems probably due to increased hygienic behavior. Lately renamed VSH aka Varroa Sensitive Hygiene.

Surplus (foundation) = Refers to thin foundation used for cut comb honey.

Surplus honey = Any extra honey removed by the beekeeper, over and above what the bees require for their own use, such as winter food stores.

Survivor stock = Bees raised from bees that were surviving without treatments. Often feral stock.

Swarm = A temporary collection of bees, containing at least one queen that split apart from the mother colony to establish a new one.

Swarm box aka a Starter hive = A box of shaken bees used to start queen cells.

Swarm cell = Queen cells usually found on the bottom of the combs before swarming.

Swarm commitment = The point just after swarm cutoff where the colony is committed to swarming.

Swarm cutoff = The point at which the colony decides to swarm or not.

Swarm trap aka Bait hive aka Decoy hive = A hive placed to attract stray swarms.

Swarm preparation = The sequence of activities of the bees that is leading up to swarming. Visually you can see this start at backfilling the brood nest so that the queen has nowhere to lay.

Swarming = The natural method of propagation of the honey bee colony.

Swarming season = The time of year, usually late spring to early summer, when swarms usually issue.

T

Tanzanian Top Bar Hive = A top bar hive with vertical sides.

Telescopic cover = A cover with a rim that hangs down all the way around it usually used with an inner cover under it.

Ten frame = A box made to take ten frames. $16^1/_4''$ wide.

Terramycin = Called oxytet in Canada and other locations. It is an antibiotic that is often used as a preventative for American and a cure for European foulbrood diseases.

Tested queen = A queen whose progeny shows she has mated with a drone of her own race and has other qualities which would make her a good colony mother. One that has been given time to prove what her qualities are.

Tergal = Pertaining to the Tergum.

Tergite = A hard overlapping plate on the dorsal portion of an arthropod that allows it to flex. Also known as sclerite.

Tergum (plural terga) = The dorsal portion of an arthropod.

Thelytoky = A type of parthenogenetic reproduction where unfertilized eggs develop into females. Usually with bees this is referring to a colony rearing a queen from a laying worker egg. This is very rare, but documented, with European honey bees. It is common with Cape Bees.

Thin surplus foundation = A comb foundation used for comb honey or chunk honey production which is thinner than that used for brood rearing. Thinner than surplus.

Thorax = The central region of an insect to which the wings and legs are attached.

Tiger striped (queen) = Markings of a particular type on a queen. Not striped like a worker (who have very even bands) but more like "flames".

Top bar = The top part of a frame or, in a top bar hive, just the piece of wood from which the comb hangs.

Top Bar Hive = a hive with only top bars and no frames that allows for movable comb without as much carpentry or expense.

Top feeder = Miller feeder. A box that goes on top of the hive that contains the syrup. See Miller Feeder.

Top supering = The act of placing honey supers on *top* of the top super of a colony as opposed to putting it under all the other supers, and directly on top of the brood box, which would be *bottom* supering or adding boxes below the brood box which would be nadiring.

Tracheal Mites = A mite that infests the trachea of the honey bee. Resistance to tracheal mites is easily bred for.

Transferring or cut out = The process of changing bees and combs from trees, houses or bee gums or skeps to movable frame hives.

Travel stains = The darkened appearance on the surface of honeycomb caused by bees walking over its surface.

Triple-wide = A box that is three times as wide as a standard ten frame box. $48^3/_4$".

Trophallaxis = The transfer of food or pheromones among members of the colony through mouth-to-mouth feeding. It is used to keep a cluster of bees alive as the edges of the cluster collect food and share it through the cluster. It is also used for communication as pheromones are shared. One very important one is QMP (Queen Mandibular Pheromone) which is shared by trophallaxis throughout the hive.

Twelve frame = A box made to take twelve frames. This is $19^7/_8$" by $19^7/_8$".

Two Queen Hive = A management method where more than one queen exists in a hive. The purpose is you get more bees and more honey with two queens.

U

Uncapping knife = A knife used to shave off the cappings of sealed honey prior to extraction; hot water, steam or electricity can heat the knives.

Uncapping tank = A container over which frames of honey are uncapped; usually strains out the honey which is then collected.

Unfertilized = An ovum or egg, which has not been united with the sperm.

Uniting = Combining two or more colonies to form a larger colony. Usually done with a sheet of newspaper between.

Unlimited Brood Nest aka "food chamber" = running bees in a configuration where the brood nest is not limited by an excluder and they are usually over-wintered in more boxes to allow more food and more expansion in the spring.

V

Varroa destructor used to be called Varroa Jacobsoni = Parasitic mite of the honey bee.

Veil = A protective netting or screen that covers the face and neck; allows ventilation, easy movement and good vision while protecting the primary targets of guard bees.

Venom allergy = A condition in which a person, when stung, may experience a variety of symptoms ranging from hives to anaphylactic shock. A person who is stung and experiences systemic (the whole body or places remote from the sting) symptoms should consult a physician before working bees again.

Venom hypersensitivity = A condition in which a person, if stung, is likely to experience an anaphylactic shock. A person with this condition should carry an emergency insect sting kit at all times during warm weather

Virgin queen = An unmated queen bee.

W

Walter T. Kelley = A beekeeping supply company out of Clarkson, KY. They have many things no one else does.

Warré hive = A type of vertical top bar hive invented by Abbé Émile Warré.

Washboarding = When the bees on the landing board or the front of a hive are moving in unison resembling a line dance.

Warming cabinet = An insulated box or room heated to liquefy honey or to heat honey to speed extraction.

Wax Dipping Hives = A method of protecting wood and also of sterilizing from AFB where the equipment is "fried" in a mixture of wax and gum resin. Usually done with paraffin sometimes done with beeswax.

Wax glands = The eight glands located on the last 4 visible, ventral abdominal segments of young worker bees; they secrete beeswax flakes.

Wax moths = See chapter *Enemies of the Bees*. Wax moths are opportunists. They take advantage of a weak hive and live on pollen, honey and burrow through the wax.

Wax scale or flake = A drop of liquid beeswax that hardens into a scale upon contact with air; in this form it is shaped into comb.

Wax tube fastener = A metal tube for applying a fine stream of melted wax to secure a sheet of foundation into a groove on a frame.

Western = I have seen this used in two ways. A box that is $6^5/_8$" in depth and the frames are $6^1/_4$" in depth. AKA Illinois AKA Medium AKA $^3/_4$ depth. Or referring to one that is $7^5/_8''$.

Western Bee Supply = A beekeeping supply company out of Montana. The company that makes all of Dadant's equipment. Also sell eight frame equipment if you request it.

Windbreaks = Specially constructed, or naturally occurring barriers to reduce the force of the (winter) winds on a beehive.

Winter cluster = A tight ball of bees within the hive to generate heat; forms when outside temperature falls below 50º F.

Winter hardiness = The ability of some strains of honey bees to survive long winters by frugal use of stored honey.

Wire, frame = Thin 28# wire used to reinforce foundation destined for the broodnest or honey extractor.

Wire cone escape = A one-way cone formed by window screen mesh used to direct bees from a house or tree into a temporary hive.

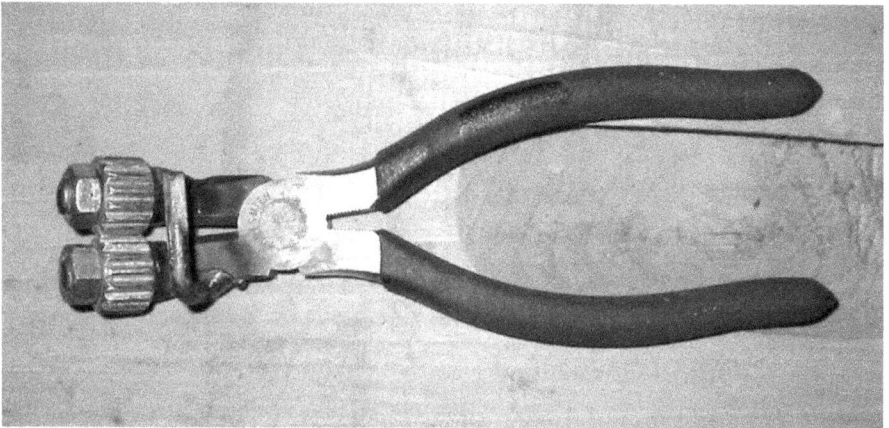

Wire crimpers = A device used to put a ripple in the frame wire to both make it tight and to distribute stress better and give more surface to bind it to the wax.

Worker bees = Infertile female bee whose re-productive organs are only partially developed, and is anatomically different than a queen and is equipped and responsible for carrying out all the routine duties of the colony.

Worker comb = Comb measuring between 4.4mm and 5.4mm, in which workers are reared and honey and pollen are stored.

Worker Queen aka laying workers = Worker bees which lay eggs in a colony hopelessly queenless; such eggs are not fertilized, since the workers cannot mate, and therefore become drones.

Worker policing = Workers that remove eggs laid by workers.

Y

Yellow (queen or bees) = When used to refer to honey bees this refers to a lighter brown color. Honey bees are *not* yellow. A Yellow queen is usually a solid light golden brown.

Acronyms

ABJ = American Bee Journal. One of the two main bee magazines in the USA.

AFB = American foulbrood

AHB = Africanized Honey Bees

AM = Apis mellifera. (European honey bees)

AMM = Apis mellifera mellifera

APV = Acute Paralysis Virus. This virus kills both adult bees and brood.

BC = Bee Culture aka Gleanings in Bee Culture. One of the two main Beekeeping magazines in the USA

BLUF = Bottom Line Up Front. A style of writing where you present the conclusion at the beginning. Common in scientific studies or military correspondence.

BPMS = Bee Parasitic Mite Syndrome

Carni = Carniolan = Apis mellifera carnica

Cauc = Caucasian = Apis mellifera Caucasia

CB = Checkerboarding (aka Nectar Management)

CCD = Colony Collapse Disorder

CPV = Chronic Paralysis Virus

CW = Conventional Wisdom

DCA = Drone Congregation Area

DVAV = Dorsal-Ventral Abdominal Vibrations dance.

DWV = Deformed Wing Virus

EAS = Eastern Apiculture Society

EFB = European Foulbrood

EHB = European Honey Bees

FGMO = Food Grade Mineral Oil

FWIW = For What It's Worth.

FWOF = Floor With Out a Floor

HAS = Heartland Apiculture Society

HBH = Honey Bee Healthy

HBTM = Honey Bee Tracheal Mite

HFCS = High Fructose Corn Syrup. A common bee feed.

HSC = Honey Super Cell (Fully drawn plastic comb in deep depth and 4.9mm cell size)

HMF = Hydroxymethyl furfural. A naturally occurring compound in honey that rises over time and rises when honey is heated.

IAPV = Israeli Acute Paralysis Virus. The virus currently being blamed for CCD

IPM = Integrated Pest Management

IMHO = In My Humble Opinion

IMO = In My Opinion

IMPOV = In My Point Of View

KTBH = Kenya Top Bar Hive (one with sloped sides)

KBV = Kashmir Bee Virus

LC = Large Cell (5.4mm cell size)

LGO = Lemon Grass (essential) Oil (used for swarm lure)

MAAREC = Mid-Atlantic Apiculture Research and Extension Consortium

NM = Nectar Management (aka Checkerboarding)

NWC = New World Carniolans

OA = Oxalic Acid. An organic acid used to kill Varroa as either a syrup or vaporized.

OSR = Oil Seed Rape (aka Canola). A crop that produces honey that is grown to produce oil.

PC = PermaComb (Fully drawn plastic comb in medium depth and about 5.0mm cell size)

PDB = Para Dichloro Benzene (aka Paramoth wax moth treatment)

PMS = parasitic mite syndrome

QMP = Queen Mandibular Pheromone

SBB = Screened Bottom Board

SBV = Sac Brood Virus

SC = Small Cell (4.9mm cell size)

SHB = Small Hive Beetle

SMR = Suppressed Mite Reproduction (usually referring to a queen)

TBH = Top Bar Hive

TM = Terramycin or Tracheal Mites depending on the context

T-Mites = Tracheal Mites

TTBH = Tanzanian Top Bar Hive (one with vertical sides)

ULBN = Unlimited Brood Nest

VD = Varroa destructor

VJ = Varroa jacobsoni

V-Mites = Varroa Mites

VSH = Varroa Sensitive Hygiene. Similar to and appears to be a more specific name for the SMR trait. A trait in queens that is being bred for where the workers sense Varroa infested cells and clean them out.

Index

About the Author

Michael Bush is one of the leading proponents of treatment free beekeeping. He has had an eclectic set of careers from printing and graphic arts, to construction to computer programming and a few more in between. Currently he is working in computers. He has been keeping bees since the mid 1970's, usually from two to seven hives up until the year 2000. Varroa forced more experimentation which required more hives and the number has grown steadily over the years from then. By 2008 it was about 200 hives. He is active on many of the Beekeeping forums with last count at about 60,000 posts between all of them. He has a web site on beekeeping at www.bushfarms.com/bees.htm

www.ingramcontent.com/pod-product-compliance
Lightning Source LLC
Chambersburg PA
CBHW021556210326
41599CB00010B/474